Groups of Divisibility

Mathematics and Its Applications (*East European Series*)

Managing Editor:

M. HAZEWINKEL

Mathematical Centre, Amsterdam, The Netherlands

Jiří Močkoř

Technical University of Mining and Metallurgy, Ostrava

Groups of Divisibility

D. REIDEL PUBLISHING COMPANY

Dordrecht / Boston / Lancaster

Library of Congress Cataloging in Publication Data

CIP

DATA APPEAR ON SEPARATE CARD

ISBN 90-277-1539-4 (REIDEL)

Published by SNTL - Publishers of Technical Literature - Prague
in co-edition with D. Reidel Publishing Company, P.O. Box 17, 3300 AA Dordrecht, Holland

Sold and distributed in Albania, Bulgaria, Chinese People's Republic, Cuba, Czechoslovakia, German Democratic Republic, Hungary, Korean People's Democratic Republic, Mongolia, Poland, Rumania, the U.S.S.R., Vietnam, and Yugoslavia by Artia, Prague, Czechoslovakia.

Sold and distributed in the U.S.A. and Canada
by Kluwer Boston, Inc.,
190 Old Derby Street, Hingham, MA 02043, U.S.A.

Distributors for all remaining countries
Kluwer Academic Publishers Group,
P.O. Box 322, 3300 AH Dordrecht, Holland.

D. Reidel Publishing Company is a member of the Kluwer Group.

All Rights Reserved.
Copyright © 1983 by SNTL - Prague, Czechoslovakia.
No part of the material protected by this copyright notice may be reproduced or utilized in any form or by any means, electronic or mechanical including photocopying, recording or by any information storage and retrieval system, without written permission from the copyright owner.

Printed in Czechoslovakia by SNTL, Prague.

CONTENTS

EDITOR'S PREFACE

INTRODUCTION 1

1. PARTIALLY ORDERED GROUPS 5
2. GROUPS OF DIVISIBILITY 13
3. BASIC PROPERTIES OF *d*-groups 22
4. *d*-CONVEX SUBGROUPS 28
5. VALUATIONS OF *d*-groups 36
6. APPROXIMATION THEOREMS 48
7. REALIZATION OF GROUPS OF DIVISIBILITY 87
8. JAFFARD-OHM's THEOREM 96
9. EXACT SEQUENCES OF GROUPS OF DIVISIBILITY 112
10. GROUPS OF DIVISIBILITY OF KRULL DOMAINS 120
11. TOPOLOGICAL GROUPS OF DIVISIBILITY 130
12. APPLICATIONS OF GROUPS OF DIVISIBILITY 149
13. GENERALIZATIONS OF SEMIVALUATIONS 163

REFERENCES 172

SUBJECT INDEX 180

EDITOR'S PREFACE

Growing specialization and diversification have brought a host of monographs and textbooks on increasingly specialized topics. However the "tree" of knowledge of mathematics and related fields does not grow only by putting forth new branches. It also happens, quite often in fact, that branches which were thought to be completely disparate are suddenly seen to be related.

Further, the kind and level of sophistication of mathematics applied in various sciences has changed drastically in recent years: measure theory is used (non-trivially) in regional and theoretical economics; algebraic geometry interacts with physics; the Minkowski lemma, coding theory and the structure of water meet one another in packing and covering theory; quantum fields, crystal defects and mathematical programming profit from homotopy theory; Lie algebras are relevant to filtering; and prediction and electrical engineering can use Stein spaces.

This program, Mathematics and Its Applications, is devoted to such (new) interrelations as exempla gratia:

- a central concept which plays an important role in several different mathematical and/or scientific specialized areas;
- new applications of the results and ideas from one area of scientific endeavor into another;
- influences which the results, problems and concepts of one field of enquiry have and have had on the development of another.

The Mathematics and Its Applications programme tries to make available a careful selection of books which fit the philosophy outlined above. With such books, which are stimulating rather than definitive, intriguing rather than encyclopaedic, we hope to contribute something towards better communication among the practitioners in diversified fields.

Because of the wealth of scholary research being undertaken in the Soviet Union, Eastern Europe, and Japan, it was decided to devote special attention to the work emanating from these particular regions.
Thus it was decided to start three regional series under the umbrella of the main MIA programme.

The present volume in the MIA (Eastern Europe) subseries is concerned with divisibility, a concept which plays an important role in several areas of mathematical inquiry, like number theory, ring theory, and the theory of ordered groups, and the applications of these fields. This book is the first which attempts to survey all aspects of divisibility and gives special emphasis to the relations between rings and partially ordered groups which are created by the procedure of assigning the appropriate group of divisibility to a ring.

Amsterdam
1983

Michiel Hazewinkel

INTRODUCTION

The theory of divisibility, the history of which is very old, covers a lot of modern algebra branches including the theory of rings, the theory of ordered groups and, of course, the theory of numbers. Thus, such a variety of applications might be the reason why a complex commentary on this branch of algebra covering the theory of divisibility as a whole has not yet been published.

At present, the theory of divisibility may be divided into two parts:

(a) Strictly multiplicative theory, and

(b) Theory of divisibility of rings.

The historical foundations of part (a) go back to the year 1930 when Arnold, Artin and Vnd der Waerden founded the theory of *ν-ideals*, although, as Aubert states in his excellent preprint [2], Dedekind had discovered ν-ideals more than half a century before they appeared in 1930; namely, that the upper half of a Dedekind cut is nothing but a ν-ideal in the ordered group of rational numbers. However, at that time the ring theoretic concept still dominated, going back to the early days of algebraic number theory, and moreover, the first works dealing with ν-ideals were still considered in the setting of rings. The first purely multiplicative conception of the theory of divisibility appears in Lorenzen's thesis in 1929. After Lorenzen's pioneer paper, a number of publication on the multiplicative theory of divisibility appeared. The most important of them was Jaffard's monography [65] and the papers of Clifford [23], [24], and Aubert [2].

The historical foundations of part (b) are much older and they go back to Kronecker's and Dedekind's first papers on ideal theory. An important contribution to this problem was Krull's papers on the theory of valuation rings which for the first time the term "group divisibility"occurs in an explicit form, From that moment on the theory of divisibility and the theory of ordered groups has been closely contacted. However, in case of part (b), this contact cannot be replaced by equality as is done-to a certain extent-in part (a) because in the theory of divisibility situations occur in which the multiplication record is not sufficient. As some of the best papers on this theme (dealing especially with the relation between the valuation rings and their value groups) the works of Ribenboim [111] - [114] may be cited.

The first works on the theory of divisibility of rings appeared together with the algebraic theory of numbers and Krull domains. The close connections between the theory of v-ideals and divisibility properties in Krull domains appeared after the so-called *theory of divisors* was introduced by Borevic and Shafarevic in [12] and reformed on a purely multiplicative notion by Skula [121]. A significant forerunner of Skula's treatment is Clifford's paper [24]. With this term is connected one of the most important results in the divisibility theory of rings due to the work of Claborn who states that a partiallyordered directed group with a special theory of divisors is a group of divisibility of a Dedekind domain.

A significant contribution to part (b) is Jaffard-Ohm's theorem which states that every (abelian) latticeordered group is a group of divisibility of a Bezout domain. This realization theorem showed a wide possibility of applying the theory of divisibility when solving the ring theoretic problems; namely, one may phrase a ring-theoretic problem in terms of ordered groups, first solve the problem there, and then pull back the solution (using some realization theorem) into the ring theory. Lorenzen [74] originally applied this technique to solve the problem of Krull and Nakayama concerning the completely integrally-closed integral domains. In the recent years this technique was successfully applied in the papers of Sheldon [117], Hill [61], Heinzer and Ohm [57], Lewis [75] and others when solving several highly interesting problems from the theory of rings.

The process given above may be converted, i.e. the results from the theory of rings may give certain results in the theory of ordered groups. One of the first applications of this process is Ohm's solution of a problem presented by Jaffard [68].

This book has grown out of efforts to write up some results which are connected with part (b) of the above division of the theory of divisibility and it is fully devoted to the invetigation of a *group of divisibility* $G(A)$ of a domain A, where $G(A)$ is the factor group $K^*/U(A)$ with K^* the multiplicative group of the quotient field of A and $U(A)$ the group of units of A with ordering defined by the positive cone $G(A)_+ =$ $= A^*/U(A)$. Contrary to the excellent paper of Aubert [2] dealing with the purely multiplicative properties of $G(A)$, we purposely keep in mind the origin of $G(A)$ from a domain A, i.e. we frequently employ properties of $G(A)$ which are not of a multiplicative nature. This access appears fully when dealing with a *d-group structure* on a group of divisibility, i.e. when we consider $G(A)$ to be a partially ordered group with a multivalued addition $\oplus_A$ which depends essentially on A. Using this so called *d*-group structure of $G(A)$ it is possible to derive a lot of properties of a domain A, using some properties of $(G(A), \oplus_A)$ even in the case where the property under the question cannot possibly be to expressed in the language of partly-ordered groups. Moreover, there is a good reason for considering such a system since it enables us to study rings and partly ordered systems in a unified way. For example, in chapter 5 is derived a theory of *Prüfer d-groups* which serves as a connection between the theory of special partly-ordered groups (e.g., lattice-ordered groups)and a theory of Prüfer domains. The part of the theory of divisibility which is connected with this system is studied in chapter 3 - 5.

In chapter 6 we first deal with approxiamation theorems for *d*-groups which enable us to obtain several theorems for valuations of fields and *l*-homomorphisms of *l*-groups. Using this method we receive, for example, a common proof of Ribenboim's approximation theorem for valuations and Krull's approximation theorem for *l*-groups. The second part of this chapter is devoted to investigating the general properties of approximation theorems from a point of view of a theory of sheaves. The notin of a representation of an approximation theorem by a sheaf is introduced and several general properties are received.

In chapter 7 we study a *realization of a group of divisibility,* i.e. a method for investigating the properties of partially-ordered groups which receive considerable attention, thus enabling us to consider a partially-ordered group as an ordered subgroup of a product of totally-ordered groups.

In chapter 8 we deal with one of the most important results in the theory of divisibility, namely, the so-called Jaffard-Ohm's theorem. In addition we show several constructions which are based on the methods used in the proofs of this theorem and we investigate the significant role which lattice-ordered groups play in the theory of divisibility.

In chapter 9 we investigate the group of divisibility of a well-known construction, the so-called composition of rings, and several others constructions which are closely connected withit. Applying the principal results of this chapter to the theory of partially ordered groups, it is possible to derive a method for constructing groups of divisibility of several ring constructions and, moreover, it enables us to construct partially ordered groups which are not groups of divisibility.

One of the most intensively investigated divisibility group is that of *Krull's domain.* We deal with this group in chapter 10, where the principal results are indicated, including Claborn's realization theorem which states that any dense ordered subgroup of a group $Z^{(I)}$ is a group of divisibility of a Dedekind domain. The results of this section are in close relation with the notion of a *theory of divisors* of a partially ordered group, i.e. a special homomorphism into a lattice ordered group which is a generalization of a theory of divisors of a Krull domain as it is known from the number theory.

Chapter 11 is devoted to investigating of the *topological groups of divisibility,* i.e. special partly ordered topological groups. We derive several realization theorems which are analogical to that of Jaffard-Ohm and, moreover, using the nonstandard methods of model theory introduced by Robinson we investigate the group of divisibility of some special completions $\hat{Z}$ of the ring of integers.

Chapter 12 we deal with some recent applications of groups of divisibility in the theory of rings and partly ordered systems. The last chapter is devoted to some notes concerning the generalization of a semivaluation.

We have tried to show full proofs of results in almost all cases presented here with the exception of several results requiring either very complicated proofs or a deeper knowledge in other branches of mathematics.

The present book may thus serve as a survey of up-to-date results which connect the three mentioned branches of algebra using the notion of divisibility and, therefore, it may be useful for specialists who are interested in any one of the three branches and who want to apply their results in the other two areas.

1. PARTIALLY-ORDERED GROUPS

In this chapter we establish some basic facts concerning partially-ordered groups, many of which are due to G. Birkhoff. Additional material may be found in the books of Fuchs [38] and Conrad [31] and in the paper by Birkhof [11.]

All groups presented in this chapter are abelian and we use multiplicative or additive notation of operation according to usage.

A group $(G, \cdot)$ is a *partially-ordered group (po-group)* if G is an order relation $\leqslant$ defined such that for every $a, b, c \in G$

$$a \leqslant b \text{ implies } ac \leqslant bc.$$

A *po*-group $(G, \cdot, \leqslant)$ is called a *lattice-ordered group* (*l-group*) if $(G, \leqslant)$ is a lattice, and it is called *totally ordered* (*o-group*) if $(G, \leqslant)$ is a totally ordered set. We shall denote the *positive cone* of G by $G_+ (= G^+)$:

$$G_+ = \{ a \in G : a \geqslant 1_G = 1 \}.$$

A *po*-group G is called *directed* if

$$\forall a, b \in G \quad \exists c \in G \text{ such that } c \leqslant a, b.$$

The positive cone of a *po*-group G has the following properties.

$$G_+ \cdot G_+ \subseteq G_+,$$
$$G_+ \cap G_+^{-1} = \{1\},$$

where $G_+^{-1} = \{ a^{-1} : a \in G_+ \}$. Conversely, if P is a subset of a group G such that

$$P \cdot P \subseteq P, \quad P \cap P^{-1} = \{1\},$$

$(G, \cdot)$ is a *po*-group with the order relation

$$a \leqslant b \text{ iff } a^{-1} \cdot b \in P \; ; \; a, b \in G,$$

and, moreover, $G_+ = P$. A *po*-group created in this way is directed if and only if $P \cdot P^{-1} = G$ (Clifford), and it is an *o*-group iff $P \cup P^{-1} = G$.

A subgroup H of a *po*-group G is called a *convex subgroup* of G

$$\forall a \in G, \; \forall b \in H, \; 1 \leqslant a \leqslant b \text{ imply } a \in H.$$

It is easy to see that an intersection of convex subgroups is a convex subgroup and that is why we may speak about a convex subgroup $C(X)$ of G generated by a subset X of G. If we denote by $[X]$ a subgroup of G generated by X, we obtain

$$C(X) = C([X]) = \{a \in G : \exists b_1, b_2 \in [X], b_1 \leq a \leq b_2\} = [X] \cdot G_+ \cap [X] . G_+^{-1} .$$

If a convex subgroup H of a *po*-group G is directed with respect to an order relation induced from G, it is called an *o-ideal* of G.

We denote by $\mathbf{C}(G)$ $(O(G))$ the set of all convex subgroups (*o*-ideals) of G.

The following theorem is of importance in the theory of *o*-groups.

1.1. THEOREM (Hölder). *For an o-group* $(G, \cdot)$ *the following are equivalent.*

(1) G *is Archimedean (i.e.* $1 < a, b \in G$ *imply* $a^n > b$ *for some positive integer* n).

(2) G *is a subgroup of a group of real numbers.*

(3) G *has no proper (i. e.* $\neq G, \{1\}$) *convex subgroups.*

If H is a convex subgroup of a *po*-group G, on the factor group G/H, we may define an order relation in the following way:

$$aH \leq bH \text{ iff } \exists c \in H \text{ such that } a \leq b \cdot c .$$

In case G is directed, G/H is directed too, and if G is an *l*-group, G/H is an *l*-group if and only if H is an *l-ideal*, i. e.

$$\forall a,b \in H \quad \text{imply} \quad a \wedge b \in H ,$$

where $a \wedge b = \inf\{a,b\}$. In this case, it is clear that $aH \wedge bH = (a \wedge b)H$, $aH \vee bH = (a \vee b)H$ hold in G/H.

Whenever we speak about a factor *po*-group G/H of a *po*-group G according to its convex subgroup H, we shall have in mind the order relation defined above. For an *l*-group G we denote by $\mathbf{L}(G)$ the set of *l*-ideals of G.

Let G and H be *po*-groups, $\delta : G \to H$ a group homomorphism. Then δ is called an *o-homomorphism* if $\delta(G_+) \subset H_+$; it is called an *o-epimorphism* if it is a surjection and $\delta(G_+) = H_+$; and it is called an *o-isomorphism* if it is an *o*-epimorphism and a group isomorphism. Here, $G \cong_o H$ will denote the fact that there exists an *o*-isomorphism between G and H. If G and H are *l*-groups, δ is called an *l-homomorphism* if

$$\delta(a \wedge b) = \delta(a) \wedge \delta(b) ; \quad a, b \in G .$$

Obviously, if δ is an *l*-homomorphism and a surjection, it is an *o*-epimorphism.

1.2. THEOREM. *A subgroup* H *of a po-group* G *is a kernel of an o-homomorphism iff* H *is a convex subgroup of* G. *If* δ *is an o-epimorphism of* G *onto a po-group* G', *then*

$$G' \cong_o G/\mathrm{Ker}\,\delta\ .$$

Further, we collect some basic properties of an l-group G which are mostly easy to be prove.

(1) $\forall a,b,c \in G,\ a(b \wedge c) = ab \wedge ac$ and dually for $\vee$.

(2) $\forall a,b \in G,\ (a \wedge b) = (a^{-1} \vee b^{-1})^{-1}$ and dually.

(3) $\forall a,b \in G,\ a \cdot b = (a \wedge b) \cdot (a \vee b)$.

(4) $\forall a,b,c \in G,\ a \wedge b = a \wedge c = 1$ implies $a \wedge (b \cdot c) = 1$.

If $a \wedge b = 1$ then we say that a and b are *disjoint.* From (4) it follows that the set of all elements disjoint from a fixed element in G_+ is a subsemigroup of G_+.

We now define for $a \in G$:

$$a^+ = a \vee 1,\ a^- = a^{-1} \vee 1,\ \ |a| = a \vee a^{-1}.$$

(5) $a \in G,\ a^+ \wedge a^{-1} = 1,\ a = a^+ \cdot (a^-)^{-1}$.

The following theorem is of great importance for our purposes.

1.3. THEOREM. *If* G *is an* l*-group and* $H \in \mathrm{L}(G)$*, then the following are equivalent.*

(1) $\forall a,b \in G,\ a,b \in G_+\backslash H$ *imply* $a \wedge b \in G_+\backslash H$.

(2) $\forall a,b \in G,\ a,b \in G_+\backslash H$ *imply* $a \wedge b > 1$.

(3) G/H *is an* o*-group*.

Proof. (1) ⇒ (2). It is clear.

(2) ⇒ (3). Let $aH, bH \in G/H$. Then for $\bar{a} = a(a \wedge b)^{-1} \geqslant 1$, $\bar{b} = b(a \wedge b)^{-1} \geqslant 1$ we have $\bar{a} \wedge \bar{b} = 1$ and $a = \bar{a}(a \wedge b)$, $b = \bar{b}(a \wedge b)$ and, according by the assumption, either $\bar{a} \in H$ or $\bar{b} \in H$. If, for example, $\bar{a} \in H$ holds, we have $aH = \bar{a}(a \wedge b)\ H = (a \wedge b)\ H \leqslant bH$. Therefore G/H is an o-group.

(3) ⇒ (1). Let $a,b \in G_+\backslash H$ and let $aH \geqslant bH > H$. Then $bH = aH \wedge bH = (a \wedge b)H$ and it follows that $a \wedge b \in G_+\backslash H$.

1.4. REMARK. An l-ideal H satisfying the conditions of 1.3 will be called a *prime* l*-ideal*. The concept of prime l-ideals is crucial in all the various methods used to represent l-groups.

If H is a prime l-ideal of on l-group G, the set $P = G_+\backslash H$ satisfies:

(1) $\forall a,b \in P,\ 1 < a \wedge b \in P$ holds;

(2) $\forall a \in G,\ b \in P,\ a \geqslant b$, imply $a \in P$;

(3) $\forall a,b \in G_+,\ a \cdot b \in P$, imply either $a \in P$ or $b \in P$.

A subset P of G_+ satisfying (1), (2), (3) is called a *prime filter* (or *t-prime ideal* by

Jaffard [65]). The set of all prime filters of G_+ will be denoted by $\mathrm{P}(G)$. A subset of G_+ which is maximal with respect to (1) will be called an *ultrafilter* of G_+.

1.5. THEOREM. *The map* $H \rightsquigarrow G_+ \setminus H$ *is a one-to-one map of the set of prime l-ideals of an l-group G onto* $\mathrm{P}(G)$ *and* $P \rightsquigarrow [G_+ \setminus P]$ *is the inverse map.*

If A is a subset of an l-group G, the set

$$A' = \{ g \in G \;:\; |a| \wedge |g| = 1,\; \forall a \in A \}$$

is called the *polar* of A.

1.6. LEMMA. $A' \in \mathrm{L}(G)$.

Proof. Let $a \in A$, $g \in G$ and let $x, y \in A'$. Then $1 \leqslant |x \cdot y^{-1}| \wedge |a| \leqslant (|x| \cdot |y| \cdot |x|) \wedge |a| \leqslant (|x| \wedge |a|) \cdot (|y| \wedge |a|) \cdot (|x| \wedge |a|) = 1$. Thus $x \cdot y^{-1} \in A'$. If $|g| \leqslant |x|$ then $1 \leqslant |g| \wedge |a| \leqslant |x| \wedge |a| = 1$ and we have $g \in A'$.

For an element $g \in G$ we write g' instead of $\{g\}'$.

1.7. THEOREM. *The map* $K \rightsquigarrow \overline{K} = \bigcup_{g \in K} g'$ *is a one-to-one map of the set of ultrafilters of an l-group G onto the set of minimal prime l-ideals of G. The inverse map is* $H \rightsquigarrow G_+ \setminus H$.

For the proof see Conrad [31].

Let G_i, $i \in J$, be a *po*-group, $G = \prod_{i \in J} G_i$, $G^* = \sum_{i \in J} G_i$, i.e. G is the set of all functions α from J into $\bigcup_{i \in J} G_i$ such that $\alpha(i) \in G_i$ for every $i \in J$ and $G^* = \{ \alpha \in G : \alpha(i) = 1$ for all but finitely many $i \in J \}$. Then $G(G^*)$ with a componentwise operation is called an *o-product* (*o-sum*) of *po*-groups G_i if

$$\alpha, \beta \in G(G^*), \quad \alpha \leqslant \beta \quad \text{iff} \quad \alpha(i) \leqslant \beta(i) \quad \text{for any } i \in J.$$

The concept of a realization has recieved a considerable attention: If G is a *po*-group, a family $\{ G_i : i \in J \}$ of *o*-groups with an *o*-isomorphism δ of G into *o*-product $\prod_{i \in J} G_i$ is called a *realization* of G if $Pr_i \cdot \delta$ is a surjection for every projection map $Pr_i : \prod_{j \in J} G_j \to G_i$. If G is an *l*-group, a realization is called an *l-realization* provided that δ is an *l*-homomorphism. The basic question of whether a given *po*-group mits a realization is completely solved in the following way:

We say that a *po*-group G is *semiclosed* if

$$\forall n \in N(\text{•}Z_+), \forall g \in G, \; g^n \geqslant 1 \text{ imply } g \geqslant 1,$$

where Z is the additive group of integers with a natural ordering. Then the following theorem holds.

1.8. THEOREM. (Lorenzen-Dieudonne). *A po-group G admits a realization if and only if it is semiclosed.*

Proof. Let G be semiclosed and let

$$M = \{P \subset G : G_+ \subset P,\ P \cdot P \subset P,\ P \cap P^{-1} = \{1\},\ P \cup P^{-1} = G\}.$$

We show that $M \neq \phi$. To do so, let

$$N' = \{P \subset G : G_+ \subset P,\ P \cdot P \subset P,\ P \cap P^{-1} = \{1\}\} \ni G_+.$$

Zorn's lemma shows the existence of a maximal element P in N'. We admit that there exists $g \in G$ such that $g \notin P \cup P^{-1}$. Then there exists $n \in N$ such that $g^n \in P^{-1}$. In fact, if for every $n \in N$, $g^n \notin P^{-1}$ holds, we set

$$Q = \{a \cdot g^n : a \in P,\ n \in N\}.$$

Then $Q \cdot Q \subset Q$, $P \subset Q$. For $x \in Q \cap Q^{-1}$ there exist $m, n \in N$, $a, b \in P$ such that $x = ag^n = b^{-1} g^{-m}$ and $x^{n+m} = a^{-1} b^{-1} \in P^{-1}$. Thus $m = n = 0$ and $x = 1$. From the maximality of P, $P = Q \ni x$ follows, a contradiction.

Analogously, for the element g^{-1} we may find an element $m \in N$ such that $g^m \in P$. Thus $g^{m \cdot n} \in P \cap P^{-1} = \{1\}$ and since G is torsion free, we obtain $g = 1$, a contradiction. Therefore $P \in M \neq \phi$.

Let $M = \{P_i : i \in J\}$ and let G_i be the o-group G with the positive cone P_i. Let

$$\delta : G \to \prod_{i \in J} G_i = \Gamma$$

be the canonical injection. It is clear that δ is an o-homomorphism. Let $g \in G_+ \cup G_+^{-1}$. Since G is semiclosed, for every $n \in N$ we have $g^n \in G_+ \cup G_+^{-1}$. Using the method analogous to the method described above it is possible to show the existence of indexes $i, j \in J$, $i \neq j$, such that $g \in P_i$, $g \in P_j^{-1}$. Thus, $\delta(g) \in \Gamma_+ \cup \Gamma_+^{-1}$ and δ is a realization of G. ∎

1.9. THEOREM (Lorenzen). *Every l-group G admits an l-realization.*

Proof. First, we prove several important lemmas.

1.10. LEMMA. *Let $H \in \mathbf{L}(G)$ and let $a \in G_+$. Then $G(H, a) = \{g \in G : |g| \leqslant a^n \cdot b$ for some $n \in N$ and $b \in H_+\}$ is the smallest l–ideal of G containing H and a. Moreover, if $a, b \in G_+$ then*

$$G(H, a) \cap G(H, b) = G(H, a \wedge b).$$

Proof. The first part is easily seen. Now, let's $1 < g \in G(H, a) \cap G(H, b)$. Then

$$g \leqslant a^n \cdot a_1,\ g \leqslant b^m \cdot b_1;\ m, n \in N,\ a_1, b_1 \in H.$$

Thus $1 < g \leqslant a^n a_1 \wedge b^m b_1 \leqslant (a \wedge b^m b_1)^n \cdot (a_1 \wedge b^m b_1) \leqslant$
$\leqslant (a \wedge b)^{m \cdot n} (a \wedge b_2)^n \cdot (a_1 \wedge b)^m \cdot (a_1 \wedge b_1) \in G(H, a \wedge b)$, and $g \in G(H, a \wedge b)$.
The other inclusion is trivial. ∎

Now, let $a \in G$, $a \neq 1$. According to a simple application of Zorn's lemma it follows that there exists an l-ideal H_a of G which is maximal without a. Then H_a is called a *value* of a.

1.11. LEMMA. *H_a is a prime l-ideal of G.*

Proof. Let $x, y \in G_+ \setminus H_a$ and assume that $x \wedge y = 1$. Then $a \in G(H_a, x) \cap$ $\cap\, G(H_a, y) = G(H_a, x \wedge y) = G(H_a, 1) = H_a$, a contradiction. Therefore, H_a is prime by 1.3. ∎

Now we return to the proof of 1.9. Let $\{H_i : i \in J\}$ be the set of prime l-ideals of G and let $a \in G$, $a \neq 1$. By 1.11., there exists $i \in J$ such that $H_i = H_a \not\ni a$, and it follows that

(1.11') $$\bigcap_{i \in J} H_i = \{1\}.$$

Let δ be the canonical map from G into o-product $\prod_{i \in J} G/H_i$. By (1.11'), δ is an l-realization. ∎

A realization $\delta : G \to \prod_{i \in J} G_i$ of a *po*-group G is called *irreducible* provided for every $i \in J$ the canonical map $G \to \prod_{i \in J, i \neq j} G_i$ is not a realization. It is possible to show that even not every l-group admits an irreducible l-realization (Jaffard [65]). On the other hand, it is possible to describe all eventually irreducible l-realizations. To do so we need some notation at first.

An element a of an l-group G is called *basic*, provided the set $\{g \in G : 1 \leqslant g \leqslant a\}$ is totally ordered. Moreover, we say that a set $S \subset G_+$ is a *basis* of G provided

(i) S is a maximal disjoint set,

(ii) each $a \in S$ is basic.

1.12. LEMMA. *If a is a basic element in an l-group G, then a' is a minimal prime l-ideal.*

Proof. Let a be basic. By 1.6., $a' \in \mathrm{L}(G)$. Let $x \cdot y \in G_+ \setminus a'$, then $a \geqslant a \wedge x$, $a \wedge y > 1$ and, for example, $a \wedge x \geqslant a \wedge y$. Then $a \wedge x = a \wedge x \wedge y = a$ and $a \leqslant x$. Thus $1 = x \wedge y \geqslant y \wedge a \geqslant 1$ and $y \in a'$, a contradiction. Therefore a' is a prime l-ideal. If there is a prime l-ideal H of G such that $H \subset a'$, $H \neq \{1\}$, there exists $g \in a'_+ \setminus H$. Then $g, a \in G_+ \setminus H$ and $g \wedge a > 1$, $g \in a'$, a contradiction. Therefore, a' is a minimal prime l-ideal of G. ∎

1.13. THEOREM. *For an l-group G the following are equivalent.*

(1) *G has an irreducible l-realization.*

(2) *G has a basis.*

If then S is a basis of G, the canonical map

$$\delta : G \to \prod_{a \in S} G/a'$$

is an irreducible l-realization.

For an easy proof see Conrad [31].

1.14. REMARK. As we have observed in the preceding consideration, for determining an l-realization δ of an l-group G it is enough (and necessery) to find a set $\{H_i : i \in J\}$ of prime l-ideals of G satisfying (1.11'). Then this set is called a *realizator* of δ.

If G and H are *po*-groups, we order the *o*-sum $G \oplus H$ by defining

$$(G \oplus H)_+ = \{(a, b) : b > 1 \text{ or } b = 1 \text{ and } a \geqslant 1\}.$$

Then the *po*-group $G \oplus H$ is called the *lexicographic sum* of G and H and it is denoted by $G \oplus_L H$.

A short exact sequence of *po*-groups

$$\{1\} \to G \xrightarrow{\alpha} H \xrightarrow{\beta} J \to \{1\} \tag{1.15}$$

is called *o-exact* if

$$\alpha(G_+) = \alpha(G) \cap H_+ , \ \beta(H_+) = J_+ .$$

Then, in particular, α and β are *o*-homomorphisms. The exact sequence (1.15) is called *lex-exact if*

$$H_+ = \{h \in H : \beta(h) > 1 \text{ or } h \in \alpha(G_+)\}.$$

It is clear that a lex-exact sequence is also *o*-exact. Then we say that a *po*-group H is an *o-extension* of a *po*-group G by a *po*-group J; or a *lex-extension of* G *by* J, depending on whether (1.15) is *o*-exact or lex-exact.

1.16. REMARK. The first construction of a lex-extension was done by Levi [73]. Special cases of lex-extension were investigated by Conrad [31], who dealt with lex-extensions of the type

$$1 \to G \to H \to H/G \to 1,$$

where G is an l-ideal of an l-group H.

His definition of lex-extension is, in this case, equivalent to the previous one which is due to Ohm [105] and Jaffard [67].

If G and H are *po*-groups, i and p are the usual injective and projective maps between G, H and $G \oplus H$, we have a lex-exact sequence

$$1 \longrightarrow G \xrightarrow{i} G \oplus_L H \xrightarrow{p} H \longrightarrow 1.$$

Further, following Ohm [105], we say that the *o*-exact sequence (1.15) *splits (splits lexico-*

graphically) if there exists a commutative diagram

$$\begin{array}{ccccccccc} 1 & \longrightarrow & G & \overset{\alpha}{\longrightarrow} & H & \overset{\beta}{\longrightarrow} & J & \longrightarrow & 1 \\ & & \| & & \downarrow \tau & & \| & & \\ 1 & \longrightarrow & G & \overset{i}{\longrightarrow} & G \oplus_L J & \overset{p}{\longrightarrow} & J & \longrightarrow & 1 \end{array}$$

where τ is an isomorphism (o-isomorphism). It is straightforward to prove that if (1.15) is lex-exact and splits, then it splits lexicographically.

1.17. LEMMA. (1) *If* (1.15) *is lex-exact and* $G, J \neq 1$, *then*

(i) H *is directed iff* J *is directed,*

(ii) H *is an* *l-group iff* G *is an l-group and* J *is an o-group.*

(2) *If* (1.15) *is o-exact and* H *is an o-group, then* G *and* J *are o-groups and* (1.15) *is lex-exact.*

2. GROUPS OF DIVISIBILITY

Let A be a commutative integral domain with the quotient field K and with the group of units $U(A)$. Let $K^* = K \setminus \{0\}$ and let

$$w_A : K^* \longrightarrow K^*/U(A) = G(A)$$

be the canonical map. The group $G(A)$ may be considered to be a directed *po*-group with the order relation

$$w_A(x) \leqslant w_A(y) \text{ iff } \exists a \in A \text{ such that } y = x \cdot a, \text{ i.e. } x \underset{A}{/} y.$$

It is clear that in this case $G(A)_+ = w_A(A \setminus \{0\})$. Then the *po*-group $G(A)$ is called a *group of divisibility* of A (or the *semivalue group of* w_A by Ohm [105], or the *value group of* A by Mott [91]. The terminology we prefer here expresses the analogy with the classical situation of division in Z. More generally, a *po*-group G is called a *group of divisibility* if there exists an integral domain A such that $G(A) \underset{o}{\cong} G$.

The map w_A may be extended on K defining

$$w_A(0) = \infty,$$

where $\infty > \alpha$ for every $\alpha \in G(A)$, $\infty + \alpha = \infty + \infty = \infty$. Then such an extended map is called a *semivaluation of a field K associated with A*.

More generally, for a field K a map w from K onto a *po*-group $(G, +)$ with an element ∞ is called a *semivaluation* provided that for every $x, y \in K$, the following holds:

(i) $w(x \cdot y) = w(x) + w(y)$,

(ii) $w(x-y) \geqslant w(z)$, for any $z \in K^*$ such that $w(z) \leqslant w(x)$, $w(y)$,
(in abbreviation : $w(x-y) \in \sup(\inf(w(x), w(y)))$),

(iii) $w(x) = \infty$ iff $x = 0$.

It is clear that every semivaluation associated with a domain satisfies (i) – (iii).

Two semivaluations w_1, w_2 with extended *po*-groups $G_1 \cup \{\infty\}$, $G_2 \cup \{\infty\}$, respectively, are called *equivalent* provided there exists an *o*-isomorphism δ such that the diagram

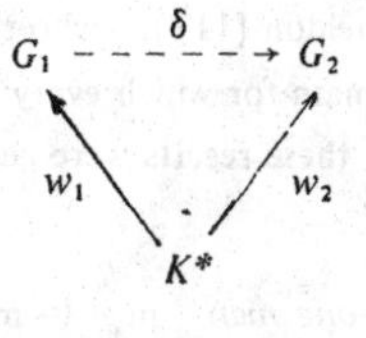

commutes. We do not distinquish equivalent semivaluations.

For a semivaluation w of a field K the set

$A_w = \{x \in K : w(x) \geqslant 0\}$

a subring of K, called the *ring of* w, and, moreover, w is equivalent to w_{A_w}. A semivaluation w of K will be called *additive* if

$w(x) < w(y)$ implies $w(x+y) = w(x)$, whenever $x+y \in K^*$.

2.1. **LEMMA.** *A semivaluation* w *of* K *is additive if and only if* A_w *is quasilocal.*

Proof. Let A_w be quasilocal. Using the equivalence of w and w_{A_w} it is easily ːen that

$M_w = \{x \in K : w(x) > 0\}$

ː the maximal ideal of A_w. If $w(x) < w(y)$, then $1 + yx^{-1} \notin M_w$ and $w(1+yx^{-1}) = 0$, $\cdot(x) = w(x+y)$. Conversely, let w be additive and let assume that A_w is not quasilocal. 'hen there exists a maximal ideal M of A_w and an element $x \in M \setminus J(A_w)$, where $J(A_w)$ ː the Jacobson radical of A_w. Then there exists $b \in A_w$ such that $1 + b\cdot x \notin U(A)$, $1+b\cdot x$ 0 and $w(b\cdot x) \geqslant w(x) > 1$, $w(bx+1) > w(1)$, a contradiction. ■

The study of divisibility of elements of a domain A amounts to the study of $;(A)$. The first observation of this was made by Krull [78], who showed the following heorem.

2.2. **THEOREM.** *A domain* R *is a valuation domain if and only if* $G(R)$ *is an -group.*

Moreover, Jaffard [67] proved that a domain A is a *GCD-domain* (or *pseudo--Bezout* by Bourbaki [13], i.e. each pair of nonzero elements of A has the greatest comnon divisor in A) iff $G(A)$ is an l-group. The proof of this fact and that of 2.2., are traightforward. Analogously, it is not difficult to show that a domain is a *UFD* iff its roup of divisibility is an o-sum of copies of Z.

Besides these straightforward facts, there is a lot of much deeper results concerning relations etween the algebraic properties of an integral domain and its group of divisibility. The first result of his type was again published by Krull [78], who showed the existence of a bijection betveen prime ideals of a valuation domain and the convex subgroups of its value group. This result vas generalized by Yakabe [135] and Sheldon [117], who showed a bijection between the prime deals of a *Bezout domain* A (i.e. a domain for which every finitely generated ideal is prinipal) and the prime filters of $G(A)_+$. All these results were generalized by Mott [92], who roved the following theorem.

2.3. **THEOREM.** *There is a one-to-one map* m_A $(= m)$ *between saturated multi-licative systems of a domain* A *and* o*-ideals of* $G(A)$. *If* S *and* H *correspond under*

$\mathbf{m}$, *then* $G(A_S) \cong_o G(A)/H$.

Proof. Let S be a saturated multiplicative system of A and let

$$\mathbf{m}(S) = [\,w_A(S)\,].$$

It is easy to see that $\mathbf{m}(S) \in O(G(A))$ and, moreover, $\mathbf{m}(S)_+ = w_A(S)$. For $H \in O(G(A))$ we set

$$\mathbf{m}^{-1}(H) = \mathbf{w}_A^{-1}(H_+).$$

Obviously, $\mathbf{m}^{-1}(H)$ is a saturated multiplicative system. Then the maps

$$H \rightsquigarrow \mathbf{m}^{-1}(H), \quad S \rightsquigarrow \mathbf{m}(S)$$

are mutually inverse. For this purpose the following must be proved:

$$H = [\,w_A(w_A^{-1}(H_+))\,], \quad S = w_A^{-1}([\,w_A(S)\,]).$$

The verification of these facts is the routine only and therefore omitted.

Finally, if S is a saturated multiplicative system of A, a map $\delta : G(A) \to G(A_S)$ defined by $\delta(w_A(x)) = w_{A_S}(x)$ is an o–epimorphism and $Ker\,\delta = \mathbf{m}(S)$. Thus, $G(A_S) \cong_o G(A)/\mathbf{m}(S)$. ∎

2.4. REMARKS. (1) P. Jaffard [65] constructed the group of divisibility of a domain A_S where S is a multiplicative system of A, and observed that $G(A_S)$ is a factor group $G(A)/H$, where

$$H = G_S \cap G_S^{-1},$$

for $G_S = G_+ \setminus w_A(S)$.

(2) Since the only saturated multiplicative systems in a valuation ring are complements of prime ideals, Krull's bijection follows directly from 2.3.

(3) To derive Yakabe and Sheldon's bijection from 2.3, by 1.5 it suffices to derive a bijection between prime ideals of A and prime l-ideals of $G(A)$. The required bijection is the restriction $\mathbf{m} \mid \{A - P : P$ a prime ideal of $A\}$. In fact, by 2.3, $G(A_P) \cong_o G(A)/\mathbf{m}(A \setminus P)$. Since A_P is a valuation domain, $G(A_P)$ is an o-group and $\mathbf{m}(A \setminus P)$ is a prime l-ideal by 1.3. Conversely, if H is a prime l-ideal of $G(A)$, $G(A_{\mathbf{m}^{-1}(H)}) \cong_o G(A)/H$ is an o-group and $A_{\mathbf{m}^{-1}(H)}$ is a valuation domain. Thus, $\mathbf{m}^{-1}(H)$ is a complement of a prime ideal of A.

(4) J. L. Mott [92] observed that the maximal ideals of a Bezout domain A correspond in Yakabe and Sheldon's bijection to the ultrafilters in $G(A)_+$.

Using 2.3., it is possible to prove the following proposition which is due to P. Sheldon [118].

2.5. PROPOSITION. *If A is a GCD-domain, then the following is equivalent.*

(1) *A is a Bezout domain*

(2) $w_A(P\setminus\{0\})$ *is a prime filter for each prime ideal* P *of* A.

(3) $w_A(M\setminus\{0\})$ *is an ultrafilter for each maximal ideal* M *of* A.

The proof may be done directly using 2.3.

As observed before, each group of divisibility is directed. On the other hand, there are directed *po*-groups which are not groups of divisibility. The first example of such a group was done by Jaffard [68] :

2.6. **EXAMPLE.** Let J be the subgroup of *o*-sum of two copies of Z such that

$(a, b) \in J$ iff $a + b$ is even,

with the induced ordering. The elementary proof of the fact that J is not a group of divisibility was done by P. Hill (see Mott [93]) :

If we assume that J is a group of divisibility of a domain A, then for certain elements $a, b \in A$,

$w_A(a) = (2, 2)$, $w_A(b) = (3, 1)$.

Then $(c, d) = w_A(a + b) \in \sup(\inf(w_A(a), w_A(b)))$ is greater than (1, 1) and (2, 0) so that $c \geqslant 2$ and $d \geqslant 1$. If $c = 2$ then $d \geqslant 2$ and if $d = 1$, then $c \geqslant 3$. In either case, $w_A(a + b) \geqslant w_A(a)$ or $w_A(a + b) \geqslant w_A(b)$. Therefore, for the principal ideals (a), (b), $(a + b)$ we have either $(a + b) \subset (a)$ or $(a + b) \subset (b)$ and it follows that either $w_A(a) \geqslant w_A(b)$ or $w_A(b) \geqslant w_A(a)$, a contradiction. ■

An unpublished result of R.L. Pendleton which was mentioned in Ohm [105] makes it possible to construct interesting examples of directed *po*-groups which are not groups of divisibility. Namely, the following theorem holds.

2.7. **THEOREM.** *The only directed orders on the group* Z *which produce groups of divisibility are the two obtained by taking either* Z_+ *or* $Z_-(= -Z_+)$ *as possitive elements.*

We prove this theorem by the help of several propositions.

2.8. **PROPOSITION**. *Let* $\succ$ *be a directed order relation on* Z *with the family* P *of positive elements. Then either* $P \subset Z_+$ *or* $P \subset Z_-$ *and* P *has only a finite number of atoms. A set* $\{m_1, \ldots, m_n\} \subset Z_+(Z_-)$ *is the set of all atoms for some directed ordering on* Z *if and only if*

(1) $GCD\ \{m_1, \ldots, m_n\} = 1$,

(2) $m_i \neq a_1 m_1 + \cdots + a_{i-1} m_{i-1} + a_{i+1} m_{i+1} + \cdots + a_n m_n$ *for every* $a_k \in Z_+(Z_-)$, $i = 1, \ldots, n$.

<u>Proof</u>. We admit that $P \not\subset Z_+$, $P \not\subset Z_-$. Then there exist $m, n \in P$ such that $m < 0$, $n > 0$. Let n_1 be the least element in Z_+ such that $n_1 \in P$, then $m + n_1 < n_1$

$m + n \in P$, a contradiction. Then we may assume that $P \subset Z_+$. Let k_p be the least element in Z_+ such that $k_p \neq 0$, $k_p \in P$. Since $\leqq$ is directed, there exists $a \in P$ such that $b = a + 1 \in P$. We consider the following family of $k_p + 1$ elements of P:

$$k_p a,\ (k_p - 1)a + b, \ldots, a + (k_p - 1)b,\ k_p b.$$

For $q \geqslant k_p b$ there exist $n, z \in Z_+$ such that

$$q = k_p \cdot n + z, \quad 0 \leqslant z < k_p .$$

Since $a k_p \leqslant a k_p + z < k_p . b$, we have $a k_p + z \in P$ and $q = (a k_p + z) + (n - a) k_p \in P$. Hence, there exists the least element n_p in Z_+ such that

$$\forall q,\ q \geqslant n_p \Rightarrow q \in P.$$

Now, if $m_1 < m_2 < \ldots < m_n < \ldots$ is an infinite family of atoms in P, there exists n such that $m_n - m \geqslant n_p$ and in this case $m_n \succ m_1$, a contradiction. Thus, in P there is only a finite number of atoms

$$m_1 < m_2 < \ldots < m_n .$$

Then

$$P = \{ \sum_{i=1}^{n} a_i m_i \ : \ a_i \in Z_+ \} .$$

In fact, let $q \in P$ and let $(a_1, \ldots, a_n) \in Z_+^n$ be a maximal element (in component wise ordering) such that

$$q \geqq \sum_{i=1}^{n} a_i m_i .$$

If $q \neq \sum_{i=1}^{n} a_i m_i$, there exists $1 \leqslant j \leqslant n$ such that $q - \sum_{i=1}^{n} a_i m_i \geqq m_j$ and in this case $(a_1, \ldots, a_n) < (a_1, \ldots, a_j + 1, \ldots, a_n)$, a contradiction. Thus, $q = \sum_{i=1}^{n} a_i m_i$. Since $1 \in P - P$, we have the $GCD\ \{m_1, \ldots, m_n\} = 1$. The rest is clear.

Conversely, let $\{m_1, \ldots, m_n\} \subset Z_+$ be a family satisfying (1), (2) and let P be as given above. Since $P + P \subset P$, $P \cap (-P) = \{0\}$, $Z = P - P$, the set P is the set of positive elements for some directed ordering on Z and using (2) it is easy to see that $m_1, \ldots, m_n$ are the atoms in P. ∎

We denote $\leqq$ a directed ordering on Z with the family $\{m_1, \ldots, m_n\} \subset Z_+$ of atoms in the positive cone P.

(2.9) *There is no proper* o*-ideal in* $(Z, \leqq)$.

In fact, let H be an o-ideal in $(Z, \leqq)$, $H \neq \{0\}$ and let $m_1, \ldots, m_t$ be all atoms of P which are contained in H. Then

$$H_+ = \{ \sum_{i=1}^{t} a_i m_i \ : \ a_i \in Z_+ \} .$$

For, for $q \in H_+ \subset P$ we have $q = \sum_{i=1}^{n} a_i m_i$. If $a_j > 0$ for $t < j \leqslant n$, then $0 \prec m_j \prec q$ and since H is convex, we have $m_j \in H$, a contradiction.

Now, let $q \in Z$, $q = \sum_{i=1}^{n} a_i m_i$, $a_i \in Z$. Then there exist $b_1, \dots, b_t \in Z_+$ such that

$$\sum_{i=1}^{t} (a_i + b_i) m_i + \sum_{j=t+1}^{n} a_j m_j \geqslant n_p ,$$

where n_p is from the proof of 2.8. Then in the factor group $(Z, \succeq)\,/\,H$ we have

$$q + H = \left(\sum_{i=1}^{t} (a_i + b_i) m_i + \sum_{j=t+1}^{n} a_j m_j\right) + H \geqslant H .$$

Therefore, $((Z, \succeq)/H)_+ = (Z, \succeq)/H$ and $H = Z$.

(2.10) *There is only a finite number of pairwise incomparable elements in* P.

In fact, let $X = \{x_1, \dots\}$ be an infinite family of pairwise incomparable elements of P, then there exist indices i, j such that $x_i > x_j + n_p$ and we have $x_i \succeq x_j$, a contradiction.

(2.11) *If* $(Z, \succeq)$ *is a group of divisibility of a domain* A *then* A *is a local ring and* $\dim A = 1$.

In fact, for an ideal J of A we set

$X = \{ w_A(x) : x \in J,\ w_A(x)$ is minimal in $w_A(J) \}$.

By 2.10, X is finite, $X = \{ w_A(x_1), \dots, w_A(x_n)\}$. Since $(P, \succeq)$ satisfies the *DCC* we have $J = (x_1, \dots, x_n)$. Hence, A is Noetherian. The rest follows from 2.3. and(2.9).

We prove the theorem. Suppose that $G = (Z, \succeq)$ is a group of divisibility of a domain A with quotient field K and let $G_+ \subset Z_+$. By(2.11), there exists the unique prime ideal M of A. Let v be a valuation of a field K with a valuation domain R_v and a maximal ideal M_v such that

$A \subset R_v$, $M_v \cap A = M$.

The existence of such a valuation follows from Gilmer [46] ; 16.5. If we denote by $\dim_A v$ the transcendental degree of the field R_v/M_v over the field A/M, it follows by Zariski, Samuel [136] ; Appendix 2, Prop. 2,

$$1 + \dim_A v \leqslant \operatorname{rank} v + \dim_A v \leqslant \dim A = 1 ,$$

hence

$$\dim_A v = 0 .$$

Thus, v is a discrete rank 1 valuation and we may assume that the value group of R_v is $(Z, \geqslant)$.

Now, for $w_A(x) \in G$ we set $\alpha(w_A(x)) = v(x)$.

Then α is an o-homomorphism of G onto $(Z, \geqslant)$. Since $\operatorname{Ker} \alpha \neq G$, we have $\operatorname{Ker} \alpha = \{0\}$

by (2.9). Let $\alpha(1) = k_0\ (\neq 0)$, then

$$\alpha(n) = \alpha(1 + \dots + 1) = k_0 \cdot n$$

and

$$v(x) = \alpha(w_A(x)) = k_0 \cdot w_A(x).$$

Since v is surjective, we have $k_0 = 1$ and

$$v(x) = w_A(x) \quad , \quad x \in K^*$$

Let $n \in Z_+$ and let $x_0 \in k$ be such that $v(x_0) = 1$.

Then

$$w_A(x_0^n + 1) = v(x_0^n + 1) = \min\{v(x_0^n), v(1)\} = 0$$

and $x_0^n + 1 = a \in A$, $x_0^n = a - 1 \in A$. Then

$$0 \prec w_A(x_0^n) = v(x_0^n) = n$$

and we have $P = Z_+$. ∎

2.12. **REMARK.** We note that using groups $(Z, \leqq)$ an example of a group of divisibility G may be constructed for which a nondirected convex subgroup H exists such that G/H *is not* a group of divisibility and such that $(G/H)_+$ has exactly n atoms.

In fact, let $\{p_1, \dots, p_n\}$ be a set of prime numbers in Z_+ and we let

$$m_i = p_1 \dots p_{i-1} p_{i+1} \dots p_n .$$

Then the family $\{m_1, \dots, m_n\}$ satisfies the conditions (1), (2) of 2.8. and there is a directed order relation $\leqq$ on Z with the atoms $m_1, \dots, m_n$ in the positive cone. We set $G = Z^n$ (the o-sum),

$$H = \{(a_1, \dots, a_n) \in G \ : \ \sum_{i=1}^{n} a_i m_i = 0\}.$$

Then H is a nondirected convex subgroup in a group of divisibility G and the map

$$G/H \longrightarrow (Z, \leqq)$$

$$(x_1, \dots, x_n) + H \longrightarrow \sum_{i=1}^{n} x_i m_i .$$

is an o-isomorphism. Hence, G/H is not a group of divisibility.

2.13. **REMARK.** We note, on the other hand, that there exists a group of divisibility G and a nondirected convex subgroup H of G such that G/H *is* a group of divisibility. An example of such a po-group has been done by T. Nakano [100].

Let $A = Z[X, Y]$ and let w be the (X, Y) – adic valuation of the quotient field K of A with the value group $(Z, \geqslant)$. Let $G = G(A)$. We set

$S = \{f \in A^* \ :$ for every irreducible polynomial p in A such that $w_A(p) \leqslant w_A(f)$ we have $w(p) \geqslant 1\}$,

$$H = \{ w_A (f \cdot g^{-1}) : f, g \in S , \ w(f) = w(g) \} .$$

Then we obtain:

(1) S is a subsemigroup of A.

(2) If $f \cdot g \in S$, then $f, g \in S$ for every $f, g \in A^*$.

(3) $U(A) = U(R_w) \cap S$.

Evidently, H is a subgroup of G. Let $w_A(f \cdot g^{-1}) = w_A(a) \in G_+ \cap H$ where $f, g \in S$, $w(f) = w(g)$. By (3), $w(a) + w(g) = w(f)$ and by (2), $a \in S$, $w(a) = w(f) - w(g) = 0$. Thus, $a \in U(A)$ and $H_+ = \{0\}$. Since $H \neq \{0\}$, H is a nondirected convex subgroup of G. We set now

$$B = A \cdot w_A^{-1}(H) \subset K .$$

We claim that B is an integral domain in K containing A. For this purpose it is sufficent to show that $A + B \subset B$. Let $a, b \in A$, $f \cdot g^{-1} \in w_A^{-1}(H)$ and let $c = ag + b.f$. Then $a + b \cdot f \cdot g^{-1} = cg^{-1}$ and

$$w(c) \geqslant \min \{ w(ag), \ w(bf) \} \geqslant w(f) = w(g) .$$

Let $n = w(c) - w(g)$. If $c = up_1 \cdots p_r \cdot q_1 \cdots q_s$ is a factorization of c into irreducible polynomials where $u = \pm 1$ and $w(p_i) = 0$, $w(q_j) \geqslant 1$, then we set $h = uq_1 \cdots q_s \in S$. Since $w(c) = w(h)$ and $w(X) = 1$, we have $w(h) = w(gX^n) = w(g) + n$ so that $h(gX^n)^{-1} \in w_A^{-1}(H)$. Hence, $cg^{-1} = p_1 \cdots p_r X^n \cdot h (gX^n)^{-1} \in B$. Thus, B is an integral domain in K.

We show that $G(B) \cong_o G/H$. At first we show that $U(B) = w_A^{-1}(H)$. To do it, let $f = af_1 g_1^{-1} \in U(B)$ where $a \in A$, $f_1 g_1^{-1} \in w_A^{-1}(H)$, $f_1, g_1 \in S$, $w(f_1) = w(g_1)$. Then $f^{-1} = b \cdot f_2 \cdot g_2^{-1} = a^{-1} \cdot f_1^{-1} g_1$, where $b \in A$, $f_2, g_2 \in S$, $w(f_2) = w(g_2)$ and we have

$$a . b = f_1^{-1} g_1 f_2^{-1} g_2 \in w_A^{-1}(H) \subset U(R_w) .$$

On the other hand, $a \cdot b \cdot f_1 \cdot f_2 = g_1 \cdot g_2 \in S$ and by (2), $a \cdot b \in S$. Thus, by (3),

$$a \cdot b \in U(R_w) \cap S = U(A)$$

and $a = \pm 1$, $f \in w_A^{-1}(H)$. The opposite inclusion is clear. If we define a map

$$\delta : G \longrightarrow G(B)$$

setting $\delta(w_A(f)) = w_B(f)$, from the fact $B = A \ U(B)$ it follows that δ is an o-epimorphism and $\operatorname{Ker} \delta = H$. Thus, $G(B) \cong_o G/H$.

An interesting method for constructing examples of *po*-groups which are not groups of divisibility provided a result for Cohen and Kaplansky [29] :

2.14. PROPOSITION. *Let G be a po-group such that it is not an l-group and let G_+ satisfies the DCC. If G is a group of divisibility, then G_+ contains more than two atoms. If G_+ contains exactly three atoms $\alpha_1, \alpha_2, \alpha_3$, then $\alpha_i + \alpha_j = 2\alpha_k$ for every* *permutation* (i, j, k) *of* 1, 2, *and* 3 .

P r o o f . Let G be a required *po*-group and let $G = G(A)$ for a domain A. Since A is not *UFD*, there exists an α_1 in G_+ such that $w_A(a_1) = \alpha_1$ for $a_1 \in A$ and, (a_1) is not a prime ideal in A. Let P be a prime ideal of A such that $a_1 \in P$. Then there exists $a_2 \in A$ such that $\alpha_2 = w_A(a_2)$ is an atom in G_+, and $a_1 + a_2 \in P$. Since G_+ satisfies the *DCC* and $\alpha_1 \leqslant w_A(a_1 + a_2)$, $\alpha_2 \leqslant w_A(a_1 + a_2)$, there exists an atom α_3, $\alpha_3 \neq \alpha_1, \alpha_2$, in G_+. Thus, G_+ has at least three atoms. Suppose G_+ has exactly three atoms $\alpha_1, \alpha_2, \alpha_3$; $\alpha_i = w_A(a_i)$, $a_i \in A$. Then $w_A(a_1 a_2 + a_3) \neq 0$, and $\alpha_1, \alpha_2 \leqslant w_A(a_1 a_2 + a_3)$; hence $\alpha_3 \leqslant w_A(a_1 a_2 + a_3)$. Thus, $a_1 a_2 = u_3 a_3^{n_3}$ where $u_3 \in U(A)$, $n_3 \in Z_+$, $n_3 \geqslant 2$. Similarly $a_1 a_3 = u_2 a_2^{n_2}$, $a_2 \,.\, a_3 = u_1 a_1^{n_1}$. Multiplying and cancelling, we obtain $n_1 = n_2 = n_3 = 2$. ∎

For example, Jaffard's example J satisfies the *DCC* on J_+ and has exactly three atoms $\alpha_1 = (1, 1)$, $\alpha_2 = (2, 0)$, $\alpha_3 = (0, 2)$. Since $\alpha_1 + \alpha_2 \neq 2\alpha_3$, J cannot be a group of divisibility.

J. Ohm [105] proved the following interesting proposition, for the proof of which see 9.6.

2.15. PROPOSITION. *Any po-group* A *is an ordered subgroup of a po-group which is not a group of divisibility.*

2.16. REMARK. The class of groups of divisibility is not a variety, since it is not closed with respect to surjective *o*-homomorphisms. On the other hand, this class is closed under ultraproducts.

In fact, let G_i, $i \in J$, be groups of divisibility, $G_i \cong_o G(A_i)$, and let $\mathbf{F}$ be a non-principal ultrafilter on J. We set

$$H = \{ f \in \prod_{i \in J} G_i \, : \, J \setminus \operatorname{supp}(f) \in \mathbf{F} \},$$

where $\operatorname{supp}(f) = \{ i \in J \, : \, f(i) \neq 0 \}$. Then H is a convex subgroup of an *o*-product $\prod_{i \in J} G_i$, and the factor group $G = \prod_{i \in J} G_i / H$ (notation : $\prod_{i \in J} G_i / \mathbf{F}$) is called an *ultraproduct* of G_i. Let B be the direct product of rings A_i, $i \in J$, and we set

$$P = \{ a \in B \, : \, J \setminus \operatorname{supp}(a) \in \mathbf{F} \}.$$

Since $\mathbf{F}$ is an ultrafilter on J, P is a prime ideal of B. Let $A = B/P$. Then $G(A) \cong_o G$.

3. BASIC PROPERTIES OF *d*-GROUPS

As. T. Nakano [100], [101] observed, there are various parallelisms between the theory of rings and that of partly ordered systems. He was especially interested in the Lorenzen's theorem concerning an *l*-realization of an *l*-group and a theorem stating that any integral domain may be expressed as an intersection of quasilocal domains. Then he showed that both the theorems may be derived from a more general theorem dealing with a special partly ordered system with a multivalued addition called a *d*-group.

The notion of a *d*-group seems to be useful not only for unifying the methods of *po*-groups and that of rings but even for the theory of groups of divisibility.

3.1. DEFINITION. A *d-group* is a *po*-group $(G, \cdot, \leqslant)$ with an element $\infty \notin G$ which admits a multivalued addition $\oplus : (G \cup \{\infty\})^2 \longrightarrow \exp(G \cup \{\infty\})$ such that

(1) $a \oplus b = b \oplus a \neq \phi$,

(2) $a \oplus (b \oplus c) = (a \oplus b) \oplus c$, where $M \oplus N = U(x \oplus y)$, $x \in M$, $y \in N$

(3) $a \in b \oplus c$ implies $b \in a \oplus c$,

(4) $a \cdot (b \oplus c) = a \cdot b \oplus a \cdot c$,

(5) $\infty \in a \oplus b$ iff $a = b$,

(6) $a, b \geqslant c$, $x \in a \oplus b$ imply $x \geqslant c$, for every $a, b, c \in G$.

The following two examples are very important for our purposes.

3.2. EXAMPLE. Let A be an integral domain, $G = G(A)$. Then for $w_A(x)$, $w_A(y)$, $w_A(z) \in G$ we define $w_A(z) \in w_A(x) \oplus_A w_A(y)$ iff there exist $u_1, u_2 \in U(A)$ such that $z = xu_1 + yu_2$.

If we set $\infty \in \alpha \oplus_A \beta$ iff $\alpha = \beta$; $\alpha, \beta \in G \cup \{\infty\}$, then $(G, \cdot, \leqslant, \oplus_A)$ is a *d*-group, called *the d-group associated with A*.

3.3. REMARK. It should be observed that the multivalued addition $\oplus_A$ on $G(A)$ is in general, dependent on A. In fact, let Z be the ring of integers, and let $A = Z_{(2)}$, $B = Z_{(3)}$. Then $G = G(A) = G(B) = (Z, +, \leqslant)$ but from the facts $0 \in 0 \oplus_B 0$, $0 \notin 0 \oplus_A 0$ it follows $\oplus_A \neq \oplus_B$.

3.4. EXAMPLE. Let G be an *l*-group. For $a, b \in G$ we set

$a \oplus_m b = \{g \in G : a \wedge b = a \wedge g = b \wedge g\}$.

If we extend $\oplus_m$ on $G \cup \{\infty\}$ by setting

$\infty \in a \oplus_m b$ iff $a = b$; $a, b \in G \cup \{\infty\}$,

then $(G, \cdot, \leqslant \oplus_m)$ is a d–group.

To show it, it suffices to prove the property (2) since the other properties are trivial. We first note, that the property (2) from 3.1 is equivalent to the axiom

(2') $(a \oplus b) \cap (c \oplus d) \neq \phi$ *implies* $(a \oplus d) \cap (b \oplus c) \neq \phi$.

In fact, for $d \in (a \oplus b) \oplus c$ there exists $x \in a \oplus b$ such that $d \in x \oplus c$. By (3), both statements $d \in (a \oplus b) \oplus c$ and $(a \oplus b) \cap (c \oplus d) \neq \phi$ are equivalent, and similarly both $d \in a \oplus (b \oplus c)$ and $(b \oplus c) \cap (a \oplus d) \neq \phi$. Hence (2') is derived from (2). Again, the axiom (2') asserts that $(a \oplus b) \oplus c \subset a \oplus (b \oplus c)$ for every a, b and c. Permuting the arrangement of a, b, c, we obtain the other inclusion.

Now, let $x \in (a \oplus_m b) \cap (c \oplus_m d)$ where $a, b, c, d \in G$. It means that

$a \wedge b = b \wedge x = a \wedge x$

and

$c \wedge d = c \wedge x = d \wedge x$.

It follows that

$a \wedge b \wedge c = b \wedge x \wedge c = b \wedge c \wedge d \leqslant a \wedge d,\ b \wedge c$.

Set $y = (a \wedge d) \wedge (b \wedge c)$. It is well known that every l- group is a distributive lattice (see Conrad [31]) and, especially, G is a modular lattice. Then by the modular law and preceding inequalities we have

$d \wedge y = (a \wedge d) \vee (b \wedge c \wedge d) = a \wedge d$,

$a \wedge y = (a \wedge d) \vee (a \wedge b \wedge c) = a \wedge d$.

So $y \in a \oplus_m d$. Similarly, we have

$b \wedge c = b \wedge y = c \wedge y$

and $y \in b \oplus_m c$. Hence, $\oplus_m$ satisfies (2'), and $(G, \oplus_m)$ is a d-group.

3.5. REMARK. If $\oplus$ is a multivalued addition on an l- group G such that $(G, \cdot, \leqslant, \oplus)$ is a d-group, then for every $a, b \in G$ we have

$a \oplus b \subset a \oplus_m b$ (in notation, $\oplus \subset \oplus_m$).

From 3.2. it follows that if a po-group G is a group of divisibility, it is possible to define a structure of a d-group on G. This enables us to find sufficient conditions for po-group G not to be a group of divisibility. For elements a, b of G we denote

$[a, b] = \{g \in G : g \leqslant a, b\}$

and the symbol $a \parallel b$ denotes that $a \nleqslant b$, $b \nleqslant a$.

3.6. PROPOSITION. *Let G be a po-group such that there exist $a, b \in G_+$ with the following properties:*

(1) *$a \wedge b$ does not exist in G,*

(2) *for every $g \in G_+$; $g \parallel a, b$, there exists $x \in [a, b]$ such that $g \parallel x$.*

Then it is not possible to define a structure of a d-group on G, and G is not a group of divisibility.

Proof. Suppose that G is a d group with respect to a multivalued addition $\oplus$. Then for $a, b = G_+$ we have

$$a \oplus b \subset \{g \in G_+ : [a, b] = [a, g] = [b, g]\}.$$

Let $a, b \in G_+$ be such that (1) and (2) hold, and let $g \in a \oplus b$. If $g \leqslant a$, we have $g \in [a, g] = [a, b]$. For $y \in [a, b]$, using (6) from 3.1., we obtain $g \geqslant y$ and $g = a \wedge b$, a contradiction. If $g > a$, we have $a \in [g, a] = [a, b]$ and again, $a = a \wedge b$. Thus, $g \parallel a$ and analogously, $g \parallel b$. Using (2), there exists $x \in [a, b]$ such that $x \parallel g$, but, on the other hand, $g \geqslant x$ us a contradiction. Thus, $a \oplus b = \phi$, and $(G, \oplus)$ is not a d-group. ∎

We note that we do not know whether $(Z, \leqq)$ satisfies the conditions of 3.6.,for every directed ordering.

3.7. EXAMPLE. If $\lesseqgtr$ is a directed ordering on Z such that the positive cone P of $(Z, \lesseqgtr)$ has exactly two atoms, then $(Z, \lesseqgtr)$ satisfies the conditions of 3.6.

In fact, let $m_1 < m_2$ be the two atoms in P and we set $a = m_1 + m_2$, $b = n.m_1$, where n is the minimal natural number such that $n.m_1 \succ m_2$. Obviously, $a \parallel b$ in $\lesseqgtr$. We suppose that there exists $a \wedge b$ in $(Z, \lesseqgtr)$. Since $m_2 \in P \cap [a, b]$, we have $a \wedge b \succ m_2$, $a \wedge b \prec m_1 + m_2$, and $m_1 + m_2 - a \wedge b \in P$, $0 < m_1 + m_2 - a \wedge b < m_1$. Since m_1 is the least element in Z_+ contained in $P - \{0\}$, we have a contradiction. Now, let $g \in P$, $g \parallel a, b$. We assume that $g \succ m_1$, $g \succ m_2$. Thus $g - m_1 \in P$, and it follows $g \succ 2m_1$. By induction we obtain $g \succ k \cdot m_1$ for every $k \in Z_+$, a contradiction. Thus, $g \parallel m_1$ or $g \parallel m_2$ and $(Z, \lesseqgtr)$ satisfies the conditions of 3.6.

3.8. REMARK. It should be observed that there exist directed *po*-groups which are not groups of divisibility and which do not satisfy the conditions of 3.6.

In fact, let $G = Z \times Z \times Z$ and we define a directed order relation on G in the following way.

$$(x, y, z) \leqslant (x_1, y_1, z_1) \text{ iff } \begin{array}{l}(x < x_1 \text{ and } y \leqslant y_1) \text{ or} \\ (x \leqslant x_1 \text{ and } y < y_1) \text{ or} \\ (x = x_1 \text{ and } y = y_1 \text{ and } z = z_1).\end{array}$$

By 9.6., G is not a group of divisibility. Suppose that $a = (g_1, a_1)$, $b = (g_2, a_2) \in G_+$

$(g_i \in Z \times Z)$ satisfy (1) and (2) of 3.6. Set $c \in Z$, $c \neq a_1, a_2$. Then $g = (g_1, c) \parallel a, b$. Condition (2) implies the existence of an element $y = (t, v) \in G_+$, such that $y \parallel g$, $y \in [a, b]$. Thus, $t \leqslant g_1 \wedge g_2 < g_1$ in $Z \times Z$ and $y < g$, a contradiction.

A d-group $(G, \oplus)$ is called *local* provided that the multivalued addition $\oplus$ is *exact*, i.e.

$a, b \in G$, $a > b$ *imply* $a \oplus b = \{b\}$.

The condition is equivalent to the axiom

$a, b \in G$, $a, b > c$, $x \in a \oplus b$ *imply* $x > c$.

In fact, let the first condition hold, and let $x \in a \oplus b$, $a, b > c$. By (6) from 3.1., $x \geqslant c$. If $x = c$, then $a \oplus x = \{x\} \not\ni b$, a contradiction. Thus, $x > c$. The converse may be done analogously.

For an l-group G, the multivalued addition $\oplus_m$ is exact if and only if G is an o-group In case G is an o-group, we may completely describe the exact multivalued additions on G.

3.9. LEMMA. *Let $(G, \oplus)$ be a totally ordered local d-group. Then either $\oplus = \oplus_m$ or $\oplus = \oplus'_m$, where for $a, b \in G$, $a \neq b$, $a \oplus'_m b = a \oplus_m b$, $a \oplus'_m a = a \oplus_m a - \{a\}$.*

P r o o f. Let $a, b, c \in G$ be such that $a \in b \oplus_m c$, $a \notin b \oplus c$. Since $\oplus$ is exact, it follows $a = b = c$, $a \notin a \oplus a$. Then for every $g \in G$ we have

$g = g \cdot a^{-1} \cdot a \notin g \cdot a^{-1} (a \oplus a) = g \oplus g$.

For every $x \in a \oplus'_m b$ we have $x = \min(a, b) < \max(a, b)$. Hence, $x \in \min(a, b) \oplus \max(a, b)$ and $\oplus = \oplus'_m$. ∎

For the d-group $(G(A), \cdot, \leqslant, \oplus_A)$ associated with a domain A we have the following result.

3.10. PROPOSITION. $(G(A), \oplus_A)$ *is a local d-group if and only if A is a quasi-local domain.*

P r o o f. Let $\oplus_A$ be exact and let

$M = \{x \in A : w_A(x) > 0\}$.

For $x, y \in M$ we have $w_A(x - y) \in w_A(x) \oplus_A w_A(y)$ and by the equivalent definition of a local d-group we have $w_A(x - y) > 0$. It is easy to see then that M is the unique maximal ideal of A.

Conversely, let A be a quasi-local then the ideal M defined above is the unique maximal ideal of A. Let $w_A(x) > w_A(y)$, $w_A(z) \in w_A(x) \oplus_A w_A(y)$. Thus, $z = xu_1 + yu_2$ for some $u_1, u_2 \in U(A) = A \setminus M$. If $w_A(z) \neq w_A(y)$, we have $w_A(z) > w_A(y)$ and $w_A(z) > w_A(y)$. Thus,

$u_2 = zy^{-1} - xy^{-1} \in M,$

a contradiction. Therefore, $w_A(z) = w_A(y)$, and $\oplus_A$ is exact. ■

3.11. **DEFINITION.** An *m-ring* is a commutative semigroup $(M, \cdot)$ with identity and an element $\infty \notin M$ that admits a multivalued addition $\oplus$, and they satisfy (1) – (15) from 3.1. All *m*-rings are required to obey the cancellation law. A subset J of an *m*-ring A is called an *m-ideal* of A provided that $a \oplus b \subset J$, $a \cdot r \in J$, for any $a, b \in J$, $r \in A$, and it is called a *prime m-ideal* if $a \cdot b \in J$ implies $a \in J$ or $b \in J$ for each $a, b \in A$.

3.12. **EXAMPLE.** Let A be a integral domain. We set

$$\overline{A} = \{\overline{x} = \{x, -x\} : x \in A\}.$$

Then $\overline{A}$ becomes an *m*-ring with respect to the multivalued addition

$$\overline{x} \oplus \overline{y} = \{\overline{x+y}, \overline{x-y}\}$$

and multiplication

$$\overline{x} \,.\, \overline{y} = \overline{x \,.\, y}\,.$$

Then $J \longleftrightarrow \overline{J} = \{\overline{x} : x \in J\}$ is an one-to-one correspondence between the set of ideal of A and the set of *m*-ideals of $\overline{A}$, and, under it, prime ideals of A correspond to prime *m*-ideals of $\overline{A}$.

Let $(A, \oplus)$ be an *m*-ring, $U(A)$ its group of units. Let $Q(A)$ be the quotient group of a semigroup $(A, \cdot)$. Then the factor group $D(A) = Q(A) / \dot{U}(A)$ is partially ordered by division with respect to A and becomes a *d*-group with respect to the multivalued addition

$$(a\,U(A))\,(b\,U(A))^{-1} \oplus' (c\,U(A))\,(d\,U(A))^{-1} =$$

$$= ((a\,d\,U(A)) \oplus (c\,b\,U(A))\,(b\,d\,U(A))^{-1}; \quad \infty = \infty\,U(A)\,.$$

Then $D(A)$ is called a *d-group related to an m-ring* A. Obviously, $D(A)$ is an analogue of a group of divisibility of a domain.

3.13. **DEFINITION.** Let $(G, \oplus)$, $(G_1, \oplus_1)$ be *d*-groups. A map δ of $G \cup \{\infty\}$ in $G_1 \cup \{\infty\}$ is called a *d-homomorphism* if δ is an *o*-homomorphism and

$$\delta(a \oplus b) \subset \delta(a) \oplus_1 \delta(b), \qquad \delta(x) = \infty \text{ iff } x = \infty\,;$$

δ is called a *d-epimorphism* if it is an *o*-epimorphism, *d*-homomorphism and for $x \in \delta(a) \oplus_1 \delta(b)$ there exist $a_1, b_1 \in G$, $y \in a_1 \oplus b_1$, such that $\delta(a_1) = a$, $\delta(b_1) = b$, $\delta(y) = x$; and δ is called a *d-isomorphism* if it is an *o*-isomorphism and

$$\delta(a \oplus b) = \delta(a) \oplus_1 \delta(b),$$

$$\delta(\infty) = \infty\,.$$

In this case we write $G \cong_d G_1$.

3.14. EXAMPLES. (1) For an integral domain A with the quotient field K we may construct two d-groups: $(G(A), \oplus_A)$ and $(D(\bar{A}), \oplus')$ where $G(A)$ is the d-group associated with A, and $D(\bar{A})$ is a d-group related to an m-ring $\bar{A}$. Their constructions are represented in the following diagram:

$$\begin{array}{ccccc} A & \longrightarrow & (G(A), \cdot, \leqslant) & \longrightarrow & (G(A), \cdot, \leqslant, \oplus_A) \\ \downarrow & & & & \downarrow \delta \\ \bar{A} & \longrightarrow & Q(\bar{A}) = \bar{K} & \longrightarrow & (D(\bar{A}), \cdot, \leqslant, \oplus') \end{array}$$

If for an element $x\,U(A) \in G(A)$ we set

$$\delta(x\,U(A)) = \bar{x}\,U(\bar{A}) \in D(\bar{A}),$$

it is easy to see that δ is a o-isomorphism and moreover,

$$\delta(x\,U(A) \oplus_A y\,U(A)) = \bar{x}\,U(\bar{A}) \oplus' \bar{y}\,U(\bar{A}) =$$
$$= \{\bar{z}\,U(\bar{A}) : \bar{z} \in \bar{x}\bar{u}_1 \oplus \bar{y}\bar{u}_2 \text{ for some } u_1, u_2 \in U(A)\}.$$

Thus, δ is an d-isomorphism, $(G(A), \oplus_A) \cong_d (D(A), \oplus')$.

(2) Let G, G_1 be l-groups. Then a map δ of G into G_1 is an l-homomorphism if and only if δ with $\delta(\infty) = \infty$ is a d-homomorphism of $(G, \oplus_m)$ into $(G_1, \oplus_m)$.

In fact, if δ is an l-homomorphism, then for $x \in a \oplus_m b$, $a, b \in G$, we have $a \wedge b = a \wedge x = b \wedge x$ and it follows $\delta(x) \in \delta(a) \oplus_m \delta(b)$, i.e. δ is a d-homomorphism. Conversely, if δ is a d-homomorphism, then since $a \wedge b \in a \oplus_m b$, and $\delta(a \wedge b) \leqslant \delta(a)$ it follows $\delta(a \wedge b) = \delta(a \wedge b) \wedge \delta(a) = \delta(a) \wedge \delta(b)$, and δ is an l-homomorphism.

4. *d*-CONVEX SUBGROUPS

Let G be a *d*-group. A subgroup H of G is called *d-convex* if it is convex and $H \cdot G_+ \oplus H \cdot G_+ \subset H \cdot G_+$. This is also equivalent to saying that if $a, b \geqslant 1$, $h \in H$ and $x \in$ $\in a \oplus b \quad h$, then there exists an element $k \in H$ such that $x \geqslant k$.

T. Nakano [100] proved the following useful lemma.

4.1. LEMMA. *Any o-ideal H of a d-group G is a d-convex subgroup of G.*

Proof. Let $a, b \geqslant 1$, $h \in H$. Then since H is directed, $h = h_1 h_2^{-1}$ where $h_i \in$ $\in H \cap G_+$. Since $a \oplus bh = (ah_2 \oplus bh_1)h_2^{-1}$ and $ah_2 \oplus bh_1 \subset G_+$, every element of $a \oplus bh$ is greater than or equal to $h_2^{-1} \in H$. Thus, H is *d*-convex. ∎

For any *d*-convex subgroup H of a *d*-group $(G, \oplus)$ it is easy to see that the factor *po*-group G/H plus the infinity element $\infty = \infty H$ becomes a *d*-group with respect to the multivalued addition

$$aH \,\tilde{\oplus}\, bH = (aH \oplus bH)/H.$$

Further, a *d*-convex subgroup H of G is called *prime* if the factor *d*-group G/H is local. One may immediately find that the primeness is originated from a consideration of the divisibility group related to a localization of an integral domain with respect to a prime ideal.

4.2. LEMMA. *Let H be a d-convex subgroup of a d-group G. If H is prime, then $a, b \geqslant 1$, $(a \oplus b) \cap H \neq \phi$ imply $a \in H$ or $b \in H$.*
Conversely, if H is directed and satisfies the condition, then it is prime.

Proof. Assume that H is prime and that neither a, nor b belongs to H. The latter means that $aH, bH > H$ in G/H. Since G/H is local and since $aH \,\tilde{\oplus}\, bH \supset (a \oplus b)H/H$, we have $xH > H$ for all $x \in a \oplus b$; this means that $(a \oplus b) \cap H = \phi$.

Next, assume that H is directed and satisfies the condition of the lemma. It suffices to prove that $aH, bH \geqslant H$, and $H \in aH \,\tilde{\oplus}\, bH$ imply $aH = H$, or $bH = H$. Further, we may assume that $a, b \geqslant 1$, and $(ah \oplus bk) \cap H \neq \phi$ for some $h, k \in H$. Since H is directed, there exists an $x \in H$ such that $h, k \geqslant x^{-1}$. Therefore, it follows that ahx, $bkx \geqslant 1$, and $(ahx \oplus bkx) \cap H \neq \phi$; these imply $aH = H$ or $bH = H$. ∎

For a *d*-group G we denote by $\mathbf{D}(G)$ ($\mathbf{M}(G)$) the set of all *d*-convex (directed prime *d*-convex) subgroups of G. The set of all directed *d*-convex subgroups of G equals $\mathbf{O}(G)$ by 4.1.

If w is a semivaluation of a field K with a value group G, then for $A = A_w$ evidently holds

$w(x-y) \in w(x) \oplus_A w(y)$.

More generaly, we say that a map w of a field K onto a d-group $(G, \oplus)$ is a *d-valuation* if for every $a, b \in K$ the following hold.

(a) $w(a \cdot b) = w(a) \cdot w(b)$,

(b) $w(a - b) \in w(a) \oplus w(b)$,

(c) $w(a) = \infty$ iff $a = 0$.

It is clear that every d-valuation is, conversely, a semivaluation with a value group G, since from the fact (b) the property (ii) from the definition of semivaluation follows using the property (6) from 3.1. By the *canonical d-valuation* of a domain A we mean a d-valuation w_A with the d-group $(G(A), \oplus_A)$.

If w_1, w_2 are d-valuations of a field K with d-groups G_1, G_2, respectively, we set $w_1 \geqslant w_2$ if there exists a d-homomorphism δ of G_1 onto G_2 such that $\delta w_1 = w_2$.

Now, if w is a d-valuation of a field K with a d-group G, for $H \in \mathbb{D}(G)$ we set

$A_w(H) = \{x \in K : w(x) \in H. G_+\}$.

Since for $x, y \in A_w(H)$, $w(x \cdot y) = w(x) \cdot w(y) \in H \cdot G_+$, $w(x-y) \in w(x) \oplus w(y) \subset \subset H \cdot G_+ \oplus H \cdot G_+ \subset H \cdot G_+$ hold, $A_w(H)$ is a subring in K, $1 \in A_w(H)$. Clearly, $A_w(\{1\}) = A_w$.

4.3. PROPOSITION. *Let G_1, G_2 be d-groups, $\delta : G_1 \longrightarrow G_2$ a d-homomorphism and let w be a d-valuation of a field K with a d-group G_1. Then δw is a d-valuation. If we suppose that $H \in \mathbb{D}(G_1)$, $Ker\ \delta \subset H$ and δ is a d-epimorphism, if follows that $\delta(H) \in \mathbb{D}(G_2)$ and $A_w(H) = A_{\delta w}(H))$.*

Proof. It is clear that $\delta.w$ is a d-valuation. Suppose that δ is a d-epimorphism and $\mathrm{Ker}\,\delta \subset H$. Then $\delta(H)$ is a convex subgroup of G_2. Let $a, b \in G_2^+$, $h_1, h_2 \in H$, then for $h \in \delta(h_1)a \oplus_2 \delta(h_2)b$ there exist $a_1, b_1 \in G_1^+$, $k_1, k_2 \in \mathrm{Ker}\,\delta \subset H$, $k \in G_1$ such that

$k \in k_1 a_1 h_1 \oplus_1 k_2 h_2 b_1 \subset H \cdot G_1^+ \oplus_1 H \cdot G_1^+ \subset H \cdot G_1^+$

and $\delta(k) = h$. Thus, $h \in \delta(H \cdot G_1^+) \subset \delta(H) \cdot G_2^+$ and $\delta(H) \in \mathbb{D}(G_2)$. The rest can be easily derived. ∎

4.4. COROLLARY. *If w is a d-valuation of a field K with a d-group G and if for $H \in \mathbb{D}(G)$ we denote the composition of w and the canonical map $G \longrightarrow G/H$ by w_H, we obtain $A_w(H) = A_{w_H}$.*

4.5. PROPOSITION. *Let w be a d-valuation of a field K with a d-group G and let $H \in \mathbb{D}(G)$, $B = A_w(H)$. Then $w_H \leqslant w_B$.*

P r o o f. Let $a = w_B(x) \in G(B)$ and let $\delta(a) = w_H(x)$. Since $A_w(H) = A_{w_H}$, for $u \in U(B)$, we have $w_H(u) = 1$ and the definition of δ is correct. For $w_B(z) \in w_B(x) \oplus_B w_B(y)$ there exist $u_1, u_2 \in U(B)$ such that $z = xu_1 + yu_2$. Hence,

$$\delta(w_B(z)) \in w_H(xu_1) \widetilde{\oplus} w_H(yu_2) = w_H(x) \widetilde{\oplus} w_H(y),$$

and $\delta(\alpha \oplus_B \beta) \subset \delta(\alpha) \widetilde{\oplus} \delta(\beta)$ where $\widetilde{\oplus}$ is the factor multivalued addition on G/H. It is now clear that δ is a d-homomorphism, and $\delta w_B = w_H$. ∎

It is easy to see that if w is a d-valuation of a field K with a d-group G and $A = A_w(H)$ for some $H \in \mathrm{D}(G)$, the addition $\oplus_A$ is the smallest one on G/H for which w_H is a d-valuation.

Further, we note that it is possible to describe all rings B in a field K such that there exists $H \in \mathrm{D}(G, \oplus_A)$ with $B = A_{w_A}(H)$. For this purpose we say that an overring B of A in a field K is *well centred* on A provided that $B = A.U(B)$, i.e. the canonical map $\delta : G(A) \longrightarrow G(B)$ defined by

$$\delta(w_A(x)) = w_B(x)$$

is an o-epimorphism.

4.6. **PROPOSITION.** *Let A be a domain with the quotient field K, and let H be a convex subgroup in $G = G(A)$. Then G/H is a group of divisibility of a domain B in K which is well centred on A if and only if $H \in \mathrm{D}(G, \oplus_A)$.*

P r o o f. Suppose that there exists a domain B such that $A \subset B \subset K$, $G(B) \cong_o G/H$, and $B = A.U(B)$. Then $H = \{w_A(x) : x \in U(B)\}$. Let $w_A(a), w_A(b) \in G_+$, $w_A(c) \in H$. For any $u_1, u_2 \in U(A)$ there exist $y \in U(B)$, $z \in A$, such that $au_1 + bcu_2 = y.z \in B$. Thus, $w_A(au_1 + bcu_2) \geqslant w_A(y) \in H$ and $H \in \mathrm{D}(G, \oplus_A)$.

Conversely, let $H \in \mathrm{D}(G, \oplus_A)$. Then $B = A_{w_H} = A_{w_A}(H)$, $A \subset B = A.U(B)$, $G(B) \cong_o G/H$. ∎

4.7. **PROPOSITION.** *Let A be a domain with the quotient field K and let $H \in \mathrm{D}(G(A), \oplus_A)$. If $\widetilde{\oplus}$ is the factor multivalued addition on $G(A)/H$, then there exists a domain B in K such that*

$$(G(B), \oplus_B) \cong_d (G(A)/H, \widetilde{\oplus}).$$

P r o o f. Let $B = A_w(H)$ where $w = w_A$. Then the map δ defined by $\delta(w_B(x)) = w_H(x)$ is an o-isomorphism between $G(B)$ and $G(A)/H$. Let $x, y, z \in K$ be such that $w_B(z) \in w_B(x) \oplus_B w_B(y)$. Then there exist $u_1, u_2 \in U(B)$ such that $z = xu_1 + yu_2$, and since $w(U(B)) = H$, we have $w(z) \in w(x)\alpha_1 \oplus_A w(y)\alpha_2$ for some $\alpha_1, \alpha_2 \in H$. Thus, $w(z)H \in w(x)H \widetilde{\oplus} w(y)H$ and since $w(x)H = \delta(w_B(x))$, we obtain

$$\delta(w_B(x) \oplus_B w_B(y)) \subset \delta w_B(x) \widetilde{\oplus} \delta w_B(y).$$

the converse inclusion is clear. ∎

4.8. PROPOSITION. *Let w be a d-valuation of a field K with a d-group $(G, \oplus)$, and let $H \in \mathrm{D}(G)$. If H is prime, $A_w(H)$ is a quasilocal domain. Conversely, if $\oplus = \oplus_A$ for $A = A_w$ and if $A_w(H)$ is quasilocal then H is prime.*

Proof. We first suppose that $H = \{1\}$ and G is a local d-group. Then $J = \{\alpha \in G : \alpha > 1\}$ is the greatest m-ideal of G_+ and it is easy to see that $M = \{x \in K : w(x) \in J\}$ is the greatest ideal of A. To consider a general case we first observe that if H is prime, G/H is a local d-group, and $A_w(H) = A_{w_H}$.

Conversely, let $\oplus = \oplus_A$ and let $B = A_w(H)$ be quasilocal. Then $(G, \oplus) = (G(A), \oplus_A)$ and $w = w_A$. By the proof of 4.7, $(G(B), \oplus_B) \cong (G(A)/H, \tilde{\oplus})$. By 3.10, $(G(B), \oplus_B)$ is a local d-group and it follows that H is prime. ∎

4.9. REMARKS (1) If G is an l-group then every l-ideal of G is directed and it follows by 4.1, it is a d-convex subgroup of $(G, \oplus_m)$. Conversely, let H be a d-convex subgroug of a d-group $(G, \oplus_m)$ and let $\alpha, \beta \in H$. Since

$$\alpha \wedge \beta \in \alpha \oplus_m \beta \subset G_+ \cdot H \oplus_m G_+ \cdot H \subset G_+ \cdot H ,$$

we have $\alpha \geqslant \alpha \wedge \beta \geqslant \rho$ for some $\rho \in H$. Since H is convex, we have $\alpha \wedge \beta \in H$. Thus, l-ideals coincide with d-convex subgroups of $(G, \oplus_m)$.

(2) For every d-valuation w of a field K with a d-group $(G, \oplus)$ the following holds

$$\mathrm{D}(G, \oplus) \subseteq \mathrm{D}(G, \oplus_A)$$

for $A = A_w$. It should be noted that, in general, the inclusion is proper. In fact, let $A = Z[X, Y]$ and let H be a nondirected convex subgroup of $G = G(A)$ constructed in 2.13. Since G/H is a group of divisibility of a domain B in the quotient field of A such that B is well centred on A, it follows by 4.6 that $H \in \mathrm{D}(G, \oplus_A)$. Since G is an l-group and H is nondirected, it is not an l-ideal and by 4.9, $H \notin \mathrm{D}(G, \oplus_m)$. Thus, $\mathrm{D}(G, \oplus_m) \subset \mathrm{D}(G, \oplus_A)$.

(3) J.L. Mott [92] introduced the notion of a prime o-ideal in a po-group G in the following way: $H \in \mathrm{O}(G)$ is *prime* if G/H is an o-group. If $G = G(A)$ for a domain A and $H \in \mathrm{O}(G)$ is prime, a ring of quotient $A_{\mathfrak{m}^{-1}(H)}$ is a valuation domain, and $\mathfrak{m}^{-1}(H) = A \setminus P$ for some prime ideal P of A. In this case we say that H is *A-prime.* Since every prime l-ideal of a group of divisibility $G(A)$ of a GCD-domain A is a prime o-ideal, it follows that every prime l-ideal of $G(A)$ is A-prime. It should be observed that in case $\oplus_A = \oplus_m$ in $G(A)$ the converse implication holds, i.e. every A-prime o-ideal of $G(A)$ is prime l-ideal and in this case A is a Bezout domain.

In fact, let H be an A-prime o-ideal of $G(A)$. Then for some prime ideal P of A, $G(A_P) \cong_o G(A)/H$. By 4.8, H is a prime d-convex subgroup of $(G(A), \oplus_A) =$

$(G(A), \oplus_m)$. Let $a, b \in G(A)_+$, $a \wedge b \in H$. Then

$a \wedge b \in (a \oplus_m b) \cap H$,

and it follows either $a \in H$ or $b \in H$. Thus, H is prime l-ideal of G. Since $G(A_P)$ is an o-group for every prime ideal P of A, A is a Bezout domain.

(4) We note that the notion of an A-prime o-ideal of $G = G(A)$ is based essentially on A. In fact, let A be a GCD-domain which is not a Bezout one. Then there exists a prime ideal P of A such that A_P is not a valuation domain, i.e. $H = \mathfrak{m}(A - P)$ is an A-prime o-ideal which is not prime l-ideal. Let B be a Bezout domain constructed in 8.1. By 8.2, $\oplus_B = \oplus_m$ and from (4) it follows that H is not B-prime even $G(B) = G(A)$ holds.

Now, we return to the investigation of general properties of d-groups.

4.10. LEMMA. *Let G be a d-group. Then there exists a one-to-one map ψ of $\mathbf{M}(G)$ onto the set of prime m-ideals of G_+ such that*

$\psi(H_1) \supset \psi(H_2)$ *iff* $H_1 \subset H_2$

for every $H_1, H_2 \in \mathbf{M}(G)$. Further, if G is directed, then for every $H \in \mathbf{M}(G)$ we have

$D((G_+)_{\psi(H)}) = G/H$

where $(G_+)_P = \{gh^{-1} : g \in G_+, h \in G_+ \setminus P\} \subset G$ is an m-ring in G and '$=$' is the equality of d-groups.

Proof. Let P be a prime m-ideal of G_+. Then the quotient subgroup $\varphi(P)$ of a semigroup $G_+ \setminus P$ is a convex directed subgroup of G and it is d-convex by 4.1. Using 4.2, it follows easily that $\varphi(P) \in \mathbf{M}(G)$. Conversely, let $H \in \mathbf{M}(G)$ and let

$\psi(H) = G_+ \setminus (H \cap G_+)$.

For $a, b \in \psi(H)$ it follows by 4.2, $a \oplus b \subseteq \psi(H)$. Since H is convex, we obtain $a \cdot b \in \psi(H)$. The condition that $\psi(H)$ is a prime m-ideal may be verified easily.

Moreover, it is easy to see that φ and ψ are mutually inverse bijections.

Suppose G is directed. Then $Q((G_+)_{\psi(H)}) = G$ and since H is an o–ideal, we have

$U((G_+)_{\psi(H)}) = \{a \cdot b^{-1} : a, b \in H \cap G_+\} = H.$

Thus, $(D =)\ D((G_+)_{\psi(H)}) = G/H$, where '$=$' means the equality between abstract groups. But, it can be seen easily that the ordering of D defined by the division with respect to $(G_+)_{\psi(H)}$ is the same as the factor ordering of G/H and the same is true for the multi-valued additions. ∎

4.11. REMARK. We note that for an l-group G the set of prime filters of G_+ coincides with the family of prime m-ideals of an m-ring $(G_+, \oplus_m)$. Then using 4.9 (1) it follows that 4.10 applied for a d-group $(G, \oplus_m)$ yields 1.5.

The our next purpose is to prove the Nakano's realization theorem for d-groups. To do it, we need several lemmas.

4.12. LEMMA. *Let* G *be a directed* d*-group,* $H \in \mathcal{O}(G)$, $p \in G$. *Then*

(1) $S = \{s \in G_+ : u \geq p, 1$ *such that* $(s \oplus u) \cap H \neq \phi\}$ *is a semigroup in* G_+.

(2) *The convex closure* $H(p)$ *of* $[S]$ *contains* H.

(3) *If* $h \geq p$, $h \in H(p)$, *then there exists* $k \in H$ *such that* $k \geq p$.

Proof. (1) Let $s, t \in S$. Then there exists $u, v \geq p, 1$ such that $(s \oplus u) \cap H \neq \phi$, and in this case, $(s \oplus u) \cdot (t \oplus v) \cap H \neq \phi$. But

$$(s \oplus u) \cdot (t \oplus v) \subset st \oplus (sv \oplus ut \oplus uv),$$

and there exists $r \in sv \oplus ut \oplus uv$ such that

$$(st \oplus r) \cap H \neq \phi, \quad st \geq 1.$$

Since $sv \geq sp \geq p, 1$; $ut \geq tp \geq p, 1$; $uv \geq up \geq p, 1$; we have $r \geq p, 1$ and $st \in S$.

(2) Let $h \in H$. Since H is directed, there exists $s \in H$ such that $s \geq h, h^{-1}, 1$. Let $v \geq p, 1$. Since $s \in s \oplus (v \oplus v)$, we have

$$s \oplus (v \oplus v) \cap H \neq \phi,$$

and there exists $u \in v \oplus v$ such that $(s \oplus u) \cap H \neq \phi$. Since $v \geq p, 1$, we have analogously $u \geq p, 1$ and it follows $s \in S$. But, $s \geq h \geq s^{-1}$ and since $H(p)$ is convex, $h \in H(p)$ holds. Thus, $H \subset H(p)$.

(3) Let $h \geq p$, $h \in H(p)$. Then there exists $s \in S$, $k \in H$, such that $s \geq h$, $k \in (s \oplus u) \cap H$ for some $u \geq p, 1$. Then $s, u \geq p$, $k \in s \oplus u$ and $k \geq p$. ∎

For an element g of a po-group G we set

$$[g) = \{x \in G : x \geq g\}.$$

4.13. LEMMA. *Let* G *be a directed* d*-group,* $p \in G$. *If* H *is a maximal* o*-ideal of* G *such that*

$$H \cap [p) = \phi,$$

then H *is a prime* d*-convex subgroup of* G.

Proof. By (2) and (3) of 4.12, we have $H = H(p)$. Let $a, b \geq 1$, $(a \oplus b) \cap H \neq \phi$ and suppose that $b \in H$. Then using the definition of $H(a)$ we obtain $v \in H(a)$ and, again by 4.12, $H \subset H(a)$. Since H is maximal with respect to the property given above, there exists $h \in H(a)$ such that $h \geq p$. Since $H(a)$ is directed, we may assume that $h \geq 1$ and $(h \oplus u) \cap H \neq \phi$ for some $u \geq a, 1$; therefore $u \in H = H(p)$. Since $u \geq a \geq 1$, we have $a \in H$ and h is prime by 4.2. ∎

4.14. THEOREM. (T. Nakano) *Let* g *be a directed* d*-group. Then*

$$\bigcap_{H \in M(G)} H = \{1\}.$$

P r o o f. Let $p \in \cap\{H : H \in M(G)\}$ and suppose that $p \neq 1$. Zorn's lemma shows then the existence of a directed d-convex subgroup H of G such that it is maximal (in the set of all o-ideals of G) in the sense that

$$H \cap [p^{-1}) = \phi .$$

By 4.13, H is prime and it follows $p^{-1} \in H$, a contradiction. Thus, $p = 1$. ∎

4.15. **REMARKS.** (1) Theorem 4.14 enables us to prove the Lorenzen's theorem 1.9 about the existence of an l-realization of an l-group using 4.9 (1) and 1.14.

(2) If G is a directed d-group, then

$$G_+ = \bigcap_{H \in M(G)} H \cdot G_+ .$$

where $H \cdot G_+$ is an m-ring in G such that $D(H \cdot G_+) = G/H$ is a local d-group. This follows directly from the proof of 4.14.

(3) Let A be an integral domain. Then using (2) for a d-group $(G(A), \oplus_A)$ we obtain

$$G(A)_+ = \bigcap_{H \in M(G(A))} H \cdot G(A)_+ ,$$

where $D(H \cdot G(A)_+) = (G(A)/H, \widetilde{\oplus})$ where $\widetilde{\oplus}$ is the factor multivalued addition on $G(A)/H$. By 4.7, $(G(A)/H, \widetilde{\oplus}) = (G(B_H), \oplus_{B_H})$ for some domain B_H. Since $G(A)/H$ is a local d-group, B_H is a quasilocal domain by 3.10. Thus,

$$G(A)_+ = \bigcap_{H \in M(G(A))} G(B_H)_+$$

and it follows

$$A = \bigcap_{H \in M(G(A))} B_H .$$

For d-groups it is possible to show an analogy of Holder's theorem 1.1. We say that a d-group G is *archimedean* provided that

$$\forall a, b \in G \quad (\forall n \in Z, a^n < b \Rightarrow a = 1).$$

For an m-ring A of a d-group G we say that A is *completely integrally closed* provided that for any $g \in G$ such that there exists $a \in G$ with the property $ag^n \in A$ for each integer $n > 0$ it follows $g \in A$.

We shall deal with the following properties of a d-group G.

(1) G is an archimedean d-group.

(2) There is no proper prime m-ideal of G_+.

(3) There is no proper prime l-convex subgroup of G.

(4) There is no proper a-convex subgroup of G.

(5) G_+ is completely integrally closed in G.

(6) If $g \in G$, $g \neq 1$, then $\bigcap_{n \in Z} (g^n \oplus g^n) = \{\infty\}$.

4.16. PROPOSITION. *Let* G *be a directed* d*-group. Then* (2) ⇔ (3) ⇔ (4), (5) ⇒ (1). *Further, if* G *is a local* d*-group then* (1) ⇔ (6) *and finally, if* G *is a totally-ordered local* d*-group, all the conditions are equivalent.*

Proof. (2) ⇒ (3). It follows by 4.10.
(3) ⇒ (4). Suppose that there is a d-convex subgroup H of G such that $H \neq \{1\}$, $H \neq G$. Then there is an element $p > 1$ such that

$$H \cap [p) = \phi$$

and the Zorn's lemma shows the existence of a d-convex subgroup H' of G maximal in the sense that $H' \cap [p) = \phi$. By 4.13, H' is a proper prime d-convex subgroup of G, a contradiction.
(4) ⇒ (2). It follows by 4.10.
(5) ⇒ (1). Suppose that $a^n < b$ for each $n \in Z$ where $a, b \in G$. Then for each $n \in Z_+$ we have $b(a^{-1})^n > 1$ and similarly, for each $n \in Z_-$, $ba^{-n} > 1$. Since G_+ is completely integrally closed, we have $a = 1$. Now, suppose that G is local.
(6) ⇒ (1). Suppose that there exist $a, b \in G$, $a \neq 1$, such that $a^n < b \neq \infty$ for each $n \in Z$. Since G is local, we obtain $a^n \oplus b = \{a^n\}$ for each $n \in Z$ and $b \in \bigcap_{n \in Z} (a^n \oplus a^n)$, a contradiction.
(1) ⇒ (6). Let $g \in G$, $g \neq 1$, and suppose that there exists $a \in G$, $a \neq \infty$, such that $a \in$ $\in \bigcap_{n \in Z} (g^n \oplus g^n)$. Then $a \geqslant g^n$ for each $n \in Z$. If we suppose $a = g^n$ for some $n \in Z$, we have $g^n \in g^{n+1} \oplus g^{n+1}$ and $1 \in g \oplus g$. Since G is local, we obtain $g = 1$, a contradiction. Thus, $a = \infty$.

Finally, we suppose G is a totally ordered local d-group and we prove (4) ⇒ (5). In fact, let $g, a \in G$ be such that $ag^n \geqslant 1$ for each $n \in Z_+$, and suppose $g < 1$. Then $a > 1$. Let H be a d-convex subgroup of G generated by $g < 1$. Since $H \neq \{1\}$, we have $G = H$ and there exists an integer m such that $1 < a^2 \leqslant g^m$. Since $g^m > 1$, it follows $m < 0$. Further, $a \geqslant g^n$ for any integer $n < 0$ and we obtain $a \geqslant g^m \geqslant a^2$, a contradiction. Thus, $g \geqslant 1$ and G_+ is a completely integrally closed. ∎

5. VALUATIONS OF *d*-GROUPS

Let G be a directed d-group, R an m-ring of G such that $G_+ \subset R$. We say that R is a *valuation m-ring* provided that $D(R)$ is a local totally ordered d-group. In this case the canonical map

$$w : G \longrightarrow G' \ (= D(R))$$

is a d-homomorphism.

More generally, every d-homomorphism w of a d-group G onto a totally ordered local d-group G will be called a *d-valuation of* G. In this case

$$R_w = w^{-1}(G'_+)$$

is clearly a valuation m-ring, and w has the following properties:

(a) $x \in a \oplus_G b$ implies $w(x) \geqslant \min\{w(a), w(b)\}$,

(b) $x \in a \oplus_G b$, $w(a) \neq w(b)$ imply $w(x) = \min\{w(a), w(b)\}$.

The main purpose here is to prove several theorems analogous to theorems for commutative rings. Doing so we need to extend the notion of an element integral over a domain to the case of d-groups at first.

Let G be a d-group. We say that an element $x \in G$ is *integral over* G_+ if there exists elements $a_0, \ldots, a_n \in G_+$ such that

$$x^{n+1} \in a_n x^n \oplus \cdots \oplus a_1 x \oplus a_0 .$$

Then G_+ is called *integrally closed* if every elements of G integral over G_+ is contained in G_+.

Moreover, for an m-ring A of a d-group G we say that A is *integrally closed* if $D(A)_+$ is integrally closed in $D(A)$.

5.1. PROPOSITION. *Let w be a d–valuation of a field K with a d-group $(G, \oplus)$ and let $H \in \mathrm{D}(G)$. If $(G/H)_+$ is integrally closed in G/H, $A_w(H)$ is integrally closed in K. The converse holds if $\oplus = \oplus_A$ for $A = A_w$.*

<u>Proof</u>. Let $x \in K$ be such that there exist $a_0, \ldots, a_n \in A_w(H)$ with $x^{n+1} = a_0 + a_1 x + \cdots + a_n x^n$. Then $w(a_i) \in H \cdot G_+$ and $w(a_i)H \geqslant H$ in G/H. Since

$$w(x)^{n+1} \in w(a_0) \oplus \cdots \oplus w(a_n) w(x)^n ,$$

we obtain

$$(w(x)H)^{n+1} \in w(a_0)H \mathbin{\tilde{\oplus}} \cdots \mathbin{\tilde{\oplus}} (w(a_n)H)(w(x)H)^n$$

and $w(x)H \geqslant H$. Thus, $x \in A_w(H)$.

Conversely, let $\oplus = \oplus_A$ and let $B = A_w(H)$ be integrally closed in K. We suppose that for $\xi \in G$ there exist $\alpha_0, \ldots, \alpha_n \in G$ such that $\alpha_i H \geqslant H$, and

$(\xi H)^{n+1} \in \alpha_0 H \tilde{\oplus} \dots \tilde{\oplus} \alpha_n \xi H$.

If we identify $G(B)$ and G/H, we obtain $\oplus_B = \tilde{\oplus}$ by 4.7, and there exist $a_0, \dots, a_n$, $a \in K$, $u_0, \dots, u_n \in U(B)$ such that

$w_H(a_i) = \alpha_i \xi^i H, \quad w_H(a) = (\xi H)^{n+1}$,

$a = a_0 u_0 + \dots + a_n u_n$.

Let $\xi = w(x)$, $\alpha_i = w(d_i)$, $w(a_i)\omega_i = \alpha_i \xi^i$ for some $\omega_i \in H = w(U(B))$; $i = 0, \dots, n$. Then there exist $v_i \in U(A)$, $y, y_i \in U(B)$, $i = 0, \dots, n$, such that

$a_i = v_i d_i x^i y_i, \quad a = x^{n+1} y$.

Thus,

$x^{n+1} = v_0 d_0 y_0 y^{-1} u_0 + \dots + v_n d_n y_n y^{-1} u_n x^n$

and since $v_i d_i y_i y^{-1} u_i \in B$, we have $x \in B$. Therefore, $w_H(x) = \xi H \geqslant H$, and $(G/H)_+$ is integrally closed. ∎

5.2. PROPOSITION. *Let $(G, \oplus)$ be a directed d-group and let $H \in O(G)$. Then if G_+ is integrally closed in G, $(G/H)_+$ is integrally closed in G/H.*

Proof. Assume that

$\xi^{n+1} H \in \alpha_n \xi^n H \tilde{\oplus} \dots \tilde{\oplus} \alpha_0 H$

where $\alpha_i H \geqslant H$ and $\tilde{\oplus}$ is the factor multivalued addition in G/H. Then there exist $\gamma^{(i)} \in H$ $(i = 0, \dots, n)$ such that

$\alpha_i \geqslant \gamma^{(i)}$; $i = 0, \dots, n$.

Further, from the definition of $\tilde{\oplus}$ it follows that there exist $\beta_0 \in \alpha_0 H, \dots, \beta_n \in \alpha_n H$ such that

(*) $\xi^{n+1} \in \xi^n \beta_n \oplus \dots \oplus \beta_0$.

Since H is directed, we can find an element $\omega \in H$ such that

$\omega \geqslant (\gamma^{(i)})^{-1}$; $i = 0, \dots, n$.

Then we have $\alpha_i \geqslant \gamma^{(i)} \geqslant \omega^{-1}$. Multiplying the relation (*) on both sides by ω^{n+1} we obtain

(**) $(\xi\omega)^{n+1} \in (\xi\omega)^n (\omega\alpha_n)\gamma_n \oplus \dots \oplus (\alpha_0 \omega)\omega^n \gamma_0$

where $\gamma_i = \beta_i \alpha_i^{-1}$. Again, since H is directed, we may find an element $\gamma \in H$ such that $\gamma \geqslant 1, (\omega^i \gamma_{n-i})^{-1}$; $i = 0, \dots, n$.

Multiplying the relation (**) on both sides by γ^{n+1}, we get

$(\xi\omega\gamma)^{n+1} \in (\xi\omega\gamma)^n \delta_n \gamma\gamma_n \oplus \dots \oplus \delta_0 \gamma^n \omega^n \gamma_0$

where $\delta_i = \alpha_i \omega \geqslant 1$, $\delta_i \gamma^{n-i+1} \omega^i \gamma_{n-i} \geqslant 1$. Since G_+ is integrally closed in G, we have

$\xi\omega\gamma \in G_+$. Thus, $\xi H \geqslant H$, and $(G/H)_+$ is integrally closed in $(G/H, \widetilde{\oplus})$. ∎

The following two lemmas are due to T. Nakano [3].

5.3. LEMMA. *Let G be a directed local d-group, $q \in G$ is not integral over G_+. Then there exist a local d-group G_q and a d-homomorphism $w : G \longrightarrow G_q$ such that* $w(q^{-1}) > 1$, $1 \notin w(G_+ - \{1\})$.

Proof. We set $p = q^{-1}$ and let

$P = \{x : x \in a_n p^n \oplus \cdots \oplus a_0,\ a_i \in G_+,\ a_0 > 1\}$,

$U = \{y : y \in b_m p^m \oplus \cdots \oplus b_1 p \oplus 1,\ b_j \in G_+\}$.

It follows easily that

i) $P \cdot P,\ P \cdot U \subset P$,

ii) $U \cdot U \subset U$,

iii) $U \oplus U \subset P \cup U$,

and since G is local,

iv) $P \oplus P \subset P$.

Again, we shall show that

v) $P \cap U = \phi$.

Suppose that

$$x \in (a_n p^n \oplus \cdots \oplus a_1 p \oplus a_0) \cap (b_m p^m \oplus \cdots \oplus b_1 p \oplus 1).$$

Then we have

$$1 \in x \oplus (b_m p^m \oplus \cdots \oplus b_1 p) \subset (a_n p^n \oplus \cdots \oplus a_1 p) \oplus (b_m p^m \oplus \cdots \oplus b_1 p) \oplus a_0$$

and it follows

$$1 \in c_s p^s \oplus \cdots \oplus c_1 p \oplus a_0$$

for some $c_i \geqslant 1$. Since $a_0 > 1$ and G is local, we obtain

$$1 \in c_s p^s \oplus \cdots \oplus c_1 p,$$

or

$$q^s \in c_s \oplus \cdots \oplus c_1 q^{s-1}$$

whence q must be integral over G_+, a contradiction. Now, set $A = (P \cup U) \cdot U^{-1}$, where $U^{-1} = \{x^{-1} : x \in U\}$. Evidently, $G_+ \subset A$, $q^{-1} \in A$. From the statements i), ii) and iii) it follows that A is an m-ring and $U \cdot U^{-1}$ is the group of units of A; further, A does not contain q. Finally, (iv) implies that $D(A)$ is a local d-group. Then the canonical d-homomorphism

$$w : G \longrightarrow G_q\ (= D(A))$$

satisfies the conditions of the lemma. ∎

5.4. LEMMA. *Let* G *be a directed local* d*-group and let every element of* G *that is not integral over* G_+ *be an inverse of an element of* G_+. *Then the integral closure* R *of* G_+ *in* G *is a valuation* m*-ring.*

Proof. Clearly

$$G = G_+^{-1} \cup R = R^{-1} \cup R .$$

Therefore, it suffices to prove that R is an m-ring of G. Suppose $p, q \in R$, and let $p \cdot q \notin R$. Then $p \cdot q \in G = G_+^{-1} \cup R$, and we have $p \cdot q < 1$. Since p is integral over G_+,

(*) $p^{n+1} \in a_n p^n \oplus \cdots \oplus a_1 p \oplus a_0 \;\; ; \; a_i \in G_+$

where the number n is the smallest one satisfying such a relation as listed above. Let $q = (ap)^{-1}$ for some $a > 1$. Since q is integral over G_+, it is

$$1 \in c_m (ap)^m \oplus \cdots \oplus c_1 (ap) \; ; \; a_i \geqslant 1 ,$$

and

$$1 \in b_m p^m \oplus \cdots \oplus b_1 p ; \; b_j \geqslant a > 1 .$$

Multiplying both sides by p^n, we have

$$p^n \in b_m p^{m+n} \oplus \cdots \oplus b_1 p^{n+1} .$$

Using (*) successively, we can descend the 'degree' of the right hand side of the relation above, until it becomes less than or equal to n. After that, we have the relation reduced such that

$$p^n \in d_n p^n \oplus \cdots \oplus d_1 p \oplus d_0 ,$$

where $d_k \geqslant a > 1$ for $k = 0, 1, \ldots, n$; since they are obtained from a_i and b_j using addition and multiplication. Since G is local, and since $d_n p^n > p^n$, it follows

$$p^n \in d_{n-1} p^{n-1} \oplus \cdots \oplus d_1 p \oplus d_0$$

and this contradicts the assumption on n.

Secondly, we need the following:

(**) If $a > 1$ and $p \in R$, then $ap > 1$.

In fact, since $R \cdot R \subset R$, $q = ap \in R$. Again, since $p \cdot q^{-1} = a^{-1} < 1$ it follows $q^{-1} \notin R$. For, for $q^{-1} \in R$ we have $a^{-1} = p \cdot q^{-1} \in R$ and for some $a_n, \ldots, a_0 \in G_+$, $b = a^{-1}$, we have

$$b^{n+1} \in a_n b^n \oplus \cdots \oplus a_1 b \oplus a_0 .$$

Since $a_i b^i \geqslant b^i \geqslant b^n$; $i = 0, \ldots, n$, it follows that $b^{n+1} \geqslant b^n$ and this implies $b = a^{-1} \geqslant 1$, a contradiction. Thus, $q^{-1} \in G = G_+^{-1} \cup R$, $q^{-1} \notin R$, and we have $q > 1$.

Finally, we shall prove that $R \oplus R \subset R$. Suppose $a^{-1} \in p \oplus q$, with $a > 1$ and $p, q \in R$ (hence, $a^{-1} \notin R$). Then $1 \in ap \oplus aq$; $aq > 1$ and this contradicts the assumption that G is local. ∎

5.5. LEMMA. *Let G be a directed d-group such that it is not local and let $S \subset G_+$ be a multiplicative closed subset with $1 \notin S$. Then there exists a prime d-convex subgroup $H \neq \{1\}$ of G such that $S \cap H = \phi$.*

Proof. Let M be a maximal m-ideal of G_+ such that $S \subset M$. It is easy to see that M is prime. By 4.10,

$H = [G_+ \setminus M] \in \mathrm{M}(G)$, $H \cap S = \phi$. ∎

For a d-valuation w of a d-group G we set

$M(w) = \{g \in G : w(g) > 1\}$.

It is clear that $M(w)$ is the unique maximal m-ideal of R_w.

The following theorem is due to J. Močkoř [82].

5.6. THEOREM. *Let G be a directed d-group, I an m-ideal of G_+. Then there exists a d-valuation w of G such that $I \subset M(w)$.*

Proof. Put

$\mathbf{X} = \{(G', w') : G'$ is a d-group, $w' : G \longrightarrow G'$ is a d-homomorphism and $1 \notin w'(I)\}$.

If we set

$(G', w') \leqslant (G'', w'')$ iff there is a d-homomorphism δ such that $\delta w' = w''$,

then it is clear that $(\mathbf{X}, \leqslant)$ is a preordered set which satisfies the conditions of Zorn's lemma; hence there exists a maximal element $(K, w) \in \mathbf{X}$. If we suppose that K is not local, by 5.5, there exists $H \in \mathrm{D}(K)$ such that for the composition w'' of w and the canonical d-homomorphism $K \longrightarrow K/H$ we have $1 \notin w''(I)$, $(K/H, w'') \in \mathbf{X}$ and $(K/H, w'') > (K, w)$; a contradiction. Thus, K is a local d-group.

We suppose that there exists $q \in K$ such that q is not integral over K_+. By 5.3, there exist a local d-group G_q and a d-homomorphism $w_q : K \longrightarrow G_q$ such that $w_q(q^{-1}) > 1$, $1 \notin w_q(K - \{1\})$. Then $1 \notin w_q w(I)$ and $(G_q, w_q w) \geqslant (K, w)$. Thus, $G_q = K$ and we may assume that w_q is an identity for every such q.

By 5.4, the integral closure R of K_+ in K is a valuation m-ring and, moreover, for the canonical map $w' : K \longrightarrow D(R)$ we have $(D(R), w'w) \geqslant (K, w)$. If we suppose that the proper inequality holds, then $1 \in w'w(I)$ and, especially, $w(I) . R = R$. In this case, let

$1 = w(u) \cdot b$; $u \in I$, $b \in R$.

Then $w(u) > 1$,

(*) $b^{n+1} \in a_n b^n \oplus \cdots \oplus a_1 b \oplus a_0$; $a_i \geqslant 1$.

Let the number n be the smallest one satisfying the relation given above. Then

$$1 = w(u)^{n+1} b^{n+1} \in w(u^{n+1}) a_n b^n \oplus \cdots \oplus w(u^{n+1}) a_0 ,$$

$$w(u^{n+1}) a_i \geqslant w(u^{n+1}) > 1 .$$

Thus,

$$1 \in c_n b^n \oplus \cdots \oplus c_1 b \oplus c_0 \quad ; \; c_i > 1 .$$

Since K is local, we obtain

$$1 \in c_n b^n \oplus \cdots \oplus c_1 b .$$

Multiplying both sides by b^n, we obtain

$$b^n \in c_n b^{2n} \oplus \cdots \oplus c_1 b^{n+1} .$$

Using (*) repeatedly, we have

$$b^n \in d_n b^n \oplus \cdots \oplus d_0$$

where $d_n > 1$. Again, thus we have

$$b^n \in d_{n-1} b^{n-1} \oplus \cdots \oplus d_0 ,$$

and this contradicts the assumption on n. Therefore, $R = K_+$ and w is the required d-valuation. ∎

5.7. COROLLARY. *Let G be a directed d-group and let P be a prime m-ideal of G_+. Then there exists a d-valuation w of G such that*

$$M(w) \cap G_+ = P .$$

P r o o f. It is easy to see that $P \cdot (G_+)_p = \{ p \cdot q^{-1} : p \in P, q \in G_+ \setminus P \} \subset G$ is the maximal m-ideal of a m-ring $(G_+)_p = \{ a \cdot b^{-1} : a \in G_+, b \in G_+ \setminus P \}$ in G. By 5.6, there exists a d-valuation w' of $D((G_+)_p)$ such that

$$w''(P.(G_+)_p) \subset M(w')$$

where

$$w'' : G \longrightarrow D((G_+)_p)$$

is the canonical d-homomorphism. Thus,

$$w''(P \cdot (G_+)_p) = M(w') \cap D((G_+)_p)_+ ,$$

and it follows

$$P(G_+)_p = M(w) \cap (G_+)_p ,$$

for $w = w'' \cdot w'$. Therefore,

$$P = P \cdot (G_+) \cap G_+ = M(w) \cap G_+ .$$

5.8. REMARKS. (1) If G is an l-group, then using the property $a \wedge b \in a \oplus_m b$ it can be easily observed that the set of m-ideals of $(G_+, \oplus_m)$ coincides with the set of filters of G_+. Then using 5.6., it is possible to derive that for any filter F of G_+ there exists an l-homomorphism δ of G onto an o-group such that $\delta(f) > 1$ for every $f \in F$.

(2) Using 5.7 it is possible to derive a well known theorem about the existence of a valuation centred on a prescribe prime ideal (see Gilmer [42], 16.5).

In fact, let A be an integral domain, P a prime ideal of A and let $G = (G(A), \oplus_A)$. Then $w_A(P)$ is a prime m-ideal of G_+ and by 5.7, there exists a d-valuation w' of G such that $M(w') \cap G_+ = w_A(P)$. We set $w = w' \cdot w_A$. Then w is a valuation of the quotient field of A, and $M(w) \cap A = P$.

(3) Let G be an o-group, then G_+ is integrally closed in $(G, \oplus_m)$. In fact, let $a_0, \ldots, a_n \in G_+$ exist for $g \in G$ such that

$$g^{n+1} \in a_n g^n \oplus_m \cdots \oplus_m a_1 g \oplus_m a_0 .$$

Then

$$g^{n+1} \geqslant a_n g^n \wedge \cdots \wedge a_1 g \wedge a_0 = a_i g^i \geqslant g^i$$

for some i, and we have $g \geqslant 1$.

(4) Let G be an l-group and let $\delta : G \longrightarrow \prod_{i \in J} G_i$ be its l-realization with $\delta_i : G \longrightarrow G_i$ the canonical l-homomorphisms. We set $A_i = \delta_i^{-1}(G_i^+)$. Then A_i is an m-ring of $(G, \oplus_m)$ and, clearly,

$$D(A_i) = (G_i, \oplus_m),\quad D(A_i)_+ = G_i^+,\quad G_+ = \bigcap_{i \in J} A_i .$$

By (3), $D(A_i)_+$ is integrally closed in $D(A_i)$ and it is easy to see that G_+ is integrally closed in G too.

5.9. LEMMA. *Let G be a directed d-group and let H be a prime d-convex subgroup of G. Then there exists a d-valuation w of G such that*

$$H \cap G_+ = U(R_w) \cap G_+ .$$

<u>Proof</u>. By 4.10., $P = G_+ \setminus H_+ = G_+ \setminus (H \cap G_+)$ is a prime m-ideal of G_+. By 5.7, there exists a d-valuation w of G such that $M(w) \cap G_+ = P$. Hence,

$$U(R_w)_+ = U(R_w) \cap G_+ = (R_w \setminus M(w)) \cap G_+ = G_+ \setminus P = H_+ . \qquad \blacksquare$$

5.10. PROPOSITION. *Let G be a directed d-group and let I be an m-ideal of G_+. Then*

$$I = \bigcap_{H \in M(G)} I.H .$$

<u>Proof</u>. We suppose $z \in I \cdot H$ for each $H \in M(G)$. Since H is directed, for any $H \in M(G)$ there exist $a_H \in I$, $h_H \in H \cap G_+$, such that $z = a_H \cdot h_H^{-1}$. Now, we put

$B = \{y \geqslant 1 : y \cdot z \in I\}$.

Clearly, B is an m-ideal of G_+ and $B \not\subset \Psi(H)$ for any $H \in \mathrm{M}(G)$. Hence, $B = G_+$ and $z \in I$. ∎

In the following we shall deal with the properties of d-groups which are analogous to these of Prüfer integral domains. To do so, we say that a d-group G is a *Prüfer d-group* provided that for every prime m-ideal P of G_+ the m-ring $(G_+)_P$ is a valuation m-ring.

Moreover, we say that a subset $F \subset G$ is a *fractional m-ideal* of G_+ provided that there exist an m-ideal A of G_+ and $g \in G$ such that $F = A \cdot g^{-1}$. An m-ideal A of G_+ is called *invertible* if there exists a fractional m-ideal F of G_+ such that $A \cdot F = G_+$.

In what follows we denote by $(a_1, \ldots, a_n)_G$ $(= (a_1, \ldots, a_n))$ an m-ideal of G_+ generated by a family $\{a_1, \ldots, a_n\} \subset G_+$, i.e.

$$(a_1, \ldots, a_n)_G = \{x : \exists g_1, \ldots, g_n \geqslant 1 \text{ such that } x \in a_1 g_1 \oplus \cdots \oplus a_n g_n\}.$$

Finally, a family $\{(G_i, \oplus_i) : i \in J\}$ of totally ordered local d-groups with a map $\delta : G \longrightarrow \prod_{i \in J} G_i$ is called a *d–realization* of a d-group $(G, \oplus)$ provided that δ is a realization of a po-group G, and

$$\delta(g \oplus h) \subset \delta(g) \oplus' \delta(h) ,$$

where

$x \in y \oplus' z$ iff $x(i) \in y(i) \oplus_i z(i)$, for every $i \in J$. We denote by $\mathrm{P}(G)$ the set of prime m-ideals of G_+.

5.11. THEOREM. *Let G be a directed d-group. Then the following conditions are equivalent.*

(1) $\{G/H : H \in \mathrm{M}(G)\}$ *is a d–realization of G.*

(2) *G in a Prüfer d-group*

(3) *G_+ is integrally closed in G, and for each m–ring A of G such that $G_+ \subset A$ there exists $\mathrm{P} \subset \mathrm{P}(G)$ such that $A = \cap (G_+)_P$ $(P \in \mathrm{P})$.*

(4) *Each m-ring A of G such that $G_+ \subset A$ is integrally closed in G.*

(5) *A factor d-group G/H is an d-group for each $H \in \mathrm{M}(G)$.*

(6) *Each finitely generated m-ideal of G_+ is invertible.*

(7) *Each m-ideal of G_+ with a basis of two elements is invertible.*

(8) *G_+ is integrally closed in G and for each $a, b \in G_+$ there exists an integer $n > 1$ such that $(a, b)^n = (a^n, b^n)$.*

(9) *G_+ is integrally closed in G and for each $a, b \in G_+$ there exists an integer $n > 1$ such that $a^{n-1} b \in (a^n, b^n)$.*

Proof. (1) ⇒ (2). Let $P \in \mathrm{P}(G)$. By 4.10, we have $D((G_+)_P) = G/\Psi^{-1}(P)$.

Since $G/\Psi^{-1}(P)$ is an o-group, it follows that G is a Prüfer d-group.

(2) ⇒ (3). Suppose that A is an m-ring of G such that $G_+ \subset A$ and let M be a prime m-ideal of A. Then $P = M \cap G_+ \in \mathrm{P}(G)$ and $(G_+)_P \subset A_M$. Hence the canonical map $D((G_+)_P) \longrightarrow D(A_M)$ is an o-homomorphism and it follows that A_M is a valuation m-ring. Since for every prime m-ideal M of $D(A)_+$ there exists a prime m-ideal M of A such that

$$D((D(A)_+)_M) \cong_d D(A_M),$$

it follows that $D(A)$ is a Prüfer d-group. By 5.10,

$$D(A)_+ = \cap H \cdot D(A)_+ \ (H \in \mathrm{M}(D(A))).$$

But, for $H \in \mathrm{M}(D(A))$, the $\Psi(H)$ is a prime m-ideal of $D(A)_+$, and $P = w_A^{-1}(\Psi(H))$ is a prime m-ideal of A where $w_A : G \longrightarrow D(A)$ is the canonical d-homomorphism. Then we obtain

$$A = \cap A_P \quad (P \text{ a prime } m\text{-ideal of } A).$$

Let $\mathrm{P}' = \{P \cap G_+ : P \text{ a prime } m \text{ ideal of } A\} \subset \mathrm{P}(G)$. For $P \cap G_+ \in \mathrm{P}'$ there exists $P' \in \mathrm{P}(G)$ such that

$$A_P = (G_+)_{P'}.$$

We denote P the family of such P'. Then

$$A = \cap (G_+)_{P'} \ (P' \in \mathrm{P}).$$

(3) ⇒ (4). Let A be an m-ring of G such that $G_+ \subset A$. Then there exists $\mathrm{P} \subset \mathrm{P}(G)$ such that $A = \cap (G_+)_P \ (P \in \mathrm{P})$. Since $D((G_+)_P) = G/\Psi^{-1}(P)$ (by 4.10) and G_+ is integrally closed in $D((G_+)_P)$, hence, $(G_+)_P$ is integrally closed. Since an intersection of integrally closed m-rings is integrally closed, we obtain A to be integrally closed.

(4) ⇒ (2). It suffices to prove that if G is a local d-group, it is totally ordered.

Let $x \in G \setminus G_+$ and let put

$$B = \{y \in G : \exists b_0, \ldots, b_m \in G_+ \text{ such that } y \in b_m x^{2m} \oplus \cdots \oplus b_1 x^2 \oplus b_0\}.$$

Clearly, B is an m-ring of G and x is integral over B; hence $x \in B$. Let

$$x \in a_0 x^{2n} \oplus \cdots \oplus a_{n-1} x^2 \oplus a_n \ ; \ a_i \geqslant 1,$$

where the number n is the smallest one satisfying the relation given above. We have

$$a_0^{2n-2}(a_0 x) \in (a_0 x)^{2n} \oplus \cdots \oplus a_n a_0^{2n-1}.$$

This means that $a_0 x$ is integral over G_+, and by the assumption that G_+ is integrally closed we obtain $a_0 x \geqslant 1$. Now, suppose that $n > 1$. Then

$$a_0^{n-1} x \in a_0^n (a_0 x^2)^n \oplus \cdots \oplus a_0^{2n-2} a_{n-1} (x a_0)^2 \oplus a_n a_0^{2n-1}$$

so that

$$a_0^{n-1} x \in (a_0 x^2)^n \oplus \cdots a_0^{n-1} a_n .$$

Since $n > 1$ we have $a_0^{n-1} x \in G_+$. Hence

$$(a_0 x^2)^n \in a_1 (a_0 x^2)^{n-1} \oplus \cdots \oplus (a_0^{n-1} a_n \oplus a_0^{n-1} x),$$

and we obtain $a_0 x^2 \geqslant 1$. Now,

$$x \in a_0 x^{2n} \oplus \cdots \oplus a_n = (a_0 x^2) x^{2n-2} \oplus \cdots \oplus a_n$$

and this contradicts the assumption on n. Therefore, $n = 1$ and $x \in a_0 x^2 \oplus a_1$ for $a_0 x \geqslant 1$. Since $a_1 \in x(1 \oplus a_0 x)$ and $x \in G_+$, we have $a_0 x = 1$ and $x^{-1} \in a_0 (a_0 x)^{-1} \geqslant$ $\geqslant 1$. Therefore, G is an o-group.

(2) ⇒ (5). Let $H \in \mathrm{M}(G)$. Since $(G_+)_{\psi(H)}$ is a valuation m-ring and $G/H = D((G_+)_{\psi(H)})$, it follows that G/H is an o-group.

(5) ⇒ (6). Let $A = (a_1, \ldots, a_n)_G$ for $a_i \in G_+$. We set

$$B = \{ g \geqslant 1 : g a_k \geqslant a_1 \text{ for } k = 1, \ldots, n \}.$$

It is easy to see that B is an m-ideal of G_+. We shall prove that

$$A \cdot B = (a_1)_G = \{ g \geqslant 1 : g \geqslant a_1 \}.$$

In fact, by 5.10, it suffices to prove that

$$(A \cdot B)/H = (a_1)/H$$

in G/H for each $H \in \mathrm{M}(G)$. First we shall prove that

$$B/H = \{ bH \geqslant H : b a_k H \geqslant a_1 H \text{ for } k = 1, \ldots, n \}.$$

Indeed, suppose that $bH \in (G/H)_+$ be such that $b a_k H \geqslant a_1 H$ for $k = 1, \ldots, n$. Then there exist $h_k \in H$; $k = 1, \ldots, n$, $h_0 \in H$, such that

$$b a_k h_k \geqslant a_1 , \quad b \geqslant h_0 \quad ; \; k = 1, \ldots, n.$$

Since H is directed, there exists $h \in H$ such that

$$h \geqslant h_k , \; h_0^{-1} \quad ; \; k = 1, \ldots, n .$$

Thus,

$$(bh) a_k \geqslant b h_k a_k \geqslant a_1 , \quad bh \geqslant 1 \; ; \; k = 1, \ldots, n .$$

Therefore, $bh \in B$ and $bH = (bh)H \in B/H$. The converse inclusion is trivial.

Now, since G/H is an o-group, for each $H \in \mathrm{M}(G)$ there exists $a_H \in \{ a_1, \ldots, a_n \}$ such that

$$A/H = (a_H H)_{G/H} .$$

Hence,

$$A \cdot B/H = \{ zgH : z a_H H \geqslant a_1 H, \; gH \geqslant a_H H \} .$$

Since $a_1 H \geqslant a_H H$, it follows that there exists $zH \geqslant H$ such that

$$a_1 H = a_H z H \geqslant a_H H$$

and we obtain

$$(a_1)_G / H \subset A . B / H .$$

The converse inclusion is trivial. Therefore, $(a_1) = A . B$ and $(B . (a_1^{-1})_G . A = G_+$. Thus, A is invertible.

(6) ⇒ (7). Trivial.

(7) ⇒ (8). It is clear that $(a, b)_G^3 = (a^3, a^2 b, a b^2, b^3)_G = (a, b)_G . (a^2, b^2)_G$. Since $(a, b)_G$ is invertible, it follows that

$$(a, b)_G^2 = (a^2, b^2)_G .$$

(8) ⇒ (9). Trivial.

(9) ⇒ (1). Let $H \in \mathrm{M}(G)$ and suppose that $gH \in G/H$. Since G is directed, there exists $a \geqslant 1$ such that $ag \geqslant 1$. Hence there exists an integer $n > 1$ such that

$$a^n g \in (a^n, (ag)^n)_G .$$

Then we have

$$a^n g \in u_1 a^n \oplus u_2 a^n g^n ,$$

for some $u_1, u_2 \geqslant 1$ and $u_1 = g u_1$ for some $u_1' \in 1 \oplus u_2 g^{n-1}$. Since G/H is a local d-group, and

$$H \in u_1' H \tilde{\oplus} u_2 g^{n-1} H ,$$

it follows that $H = u_1' H$ or $H = u_2 g^{n-1} H$. In the first case we have $H \leqslant u_1 H = g u_1' H = gH$; in the second case we have $(g^{-1})^{n-1} H = u_2 H \geqslant H$. Suppose $(g^{-1})^{n-1} H > H$. Then

$$(g^{-1})^{n-1} H \tilde{\oplus} H = \{H\} .$$

Thus, $g^{-1} H$ is integral over $(G/H)_+$ and since G_+ is integrally closed, we have $g^{-1} H \geqslant H$ by 5.2. Suppose $(g^{-1})^{n-1} H = H$. Again, $g^{-1} H$ is integrally closed over $(G/H)_+$ and $gH \leqslant H$. Therefore, G/H is an o-group and by 4.14, the canonical d-homomorphism

$$\delta : G \longrightarrow \Pi G/H \quad (H \in \mathrm{M}(G))$$

is a d-realization of G. ∎

5.12. **REMARKS.** (1) If G is an l-group, then $(G, \oplus_m)$ is a Prüfer d-group. This follows directly from the fact that every $H \in \mathrm{M}(G)$ is a prime l-ideal of G.

(2) For an integral domain A, A is a Prüfer domain if and only if $(G(A), \oplus_A)$ is a Prüfer d-group. Moreover, we can prove this more general result:

5.12.1. **PROPOSITION.** *Let w be a d-valuation of a field K with a d-group $(G, \oplus)$ and let $H \in \mathrm{D}(G)$. If $A_w(H)$ is a Prüfer domain, then G/H is a Prüfer d-group. The converse holds if $\oplus = \oplus_A$ for $A = A_w$.*

P r o o f. At first, let $H = \{1\}$ and let $\widetilde{H} \in M(G, \oplus) = M(G/H, \widetilde{\oplus})$. Then for $\alpha, \gamma \in G$ we have $\alpha \oplus_A \beta \subset \alpha \oplus \beta$ and by 4.8, $\widetilde{H} \in M(G, \oplus_A)$. For $P = w^{-1}(\Psi(\widetilde{H})) = w^{-1}(G_+ \setminus \widetilde{H})$ it is easy to see that

$$G(A_P) \cong_o G/\widetilde{H}$$
$$x\, U(A_P) \longrightarrow w(x)\widetilde{H}.$$

Thus, $G/\widetilde{H}$ is an o-group and $G(= G/H)$ is a Prüfer d-group.

Now, we consider the general case. According 4.4, we have $A_w(H) = A_{w_H}$, and w_H is a d-valuation of K with d-group G/H. According the first part, G/H is a Prüfer d-group.

Conversely, let $\oplus = \oplus_A$ and let $H \in D(G)$ be such that $(G/H, \widetilde{\oplus})$ is a Prüfer d-group. By 4.7,

$$(G(B), \oplus_B) \cong_d (G/H, \widetilde{\oplus})$$

where $B = A_w(H)$. Let M be a maximal ideal of B and let $\widetilde{H} \in O(G(B))$ be such that $G(B_M) = (G/H)/\widetilde{H}$. If we identify $G(B)$ and G/H, we have $\widetilde{\oplus} = \oplus_B$. Since B_M is quasilocal, $\widetilde{H} \in M(G/H, \widetilde{\oplus})$ by 4.8. Since G/H is a Prüfer d-group, $(G/H)/\widetilde{H}$ is an o-group and B_M is a valuation domain. Hence, B is a Prüfer domain. ∎

(3) It should be observed that in case $\oplus \neq \oplus_A$ the converse implication in 5.12.1 need not hold in general. In fact, let A be a GCD – domain that is not Bezout, i.e., A is not a Prüfer domain. Let $G = (G(A), \oplus_m)$ and let $w = w_A$. Since

$$w(x+y) \in w(x) \oplus_A w(y) \subset w(x) \oplus_m w(y),$$

w is a d-valuation with a d-group G. According (1), G is a Prüfer d-group but A is not a Prüfer domain.

On the other hand, the condition $\oplus = \oplus_A$ is not necessary in the converse implication in 5.12.1. In fact, let w be a d-valuation of the field Q of rational numbers with a d-group $G = (Z, \oplus_m)$ and let $A = Z_{(2)}$. Since $D(G) = \{\{0\}\}$, $A_w(H)$ is a Prüfer domain, but $\oplus_m \neq \oplus_{A_w}$.

6. APPROXIMATION THEOREMS

In this chapter we deal with approximation theorems for d-groups, especially, we show that d-groups mostly satisfy the same approximation theorems as the valuations of fields and l-groups.

Using such approximation theorems it is possible to obtain several others for valuations of fields and, moreover, several propositions concerning l-groups, for example, a proof of Krull's conjecture for l-groups.

For a directed d-group G we set

$W(G) = \{(G', w) \; : \; w : G \longrightarrow G' \text{ is a } d\text{-valuation of } G\}$

and we set $(G_1, w_1) \leqslant (G_2, w_2)$ if there exists a d-homomorphism $\delta : G_2 \longrightarrow G_1$ such that $\delta w_2 = w_1$. Then $(W(G), \leqslant)$ is a preordered set.

6.1. LEMMA. $(W(G), \leqslant)$ *is an inf-semilattice.*

Proof. Let $(G_1, w_1), (G_2, w_2) \in W(G)$ and let $G'' = G/(\operatorname{Ker} w_1 \cdot \operatorname{Ker} w_2)$ be an abstract factor group. We define a preordered relation $\leqslant$ on G'' as follows :

$w'(g) \leqslant w'(t)$ iff $w_i(g) \leqslant w_i(t)$; $i = 1, 2$, or
$w_1(g) < w_1(t)$, $w_2(g) > w_2(t)$, or
$w_2(g) < w_2(t)$, $w_1(g) > w_1(t)$;

where w' is the canonical map of G onto G''. Then $(G'', \leqslant)$ is a preordered group. Let $(G', \leqslant)$ be an associated po-group and let w be the composition of w' with the canonical map of G'' onto G', i.e. $w(g)$ is an element of G' containing $w'(g)$. If we denote by δ_i', δ_i, respectively, the maps defined by

$\delta_i'(w_i(g)) = w'(g)$, $\delta_i(w_i(g)) = w(g)$,

we obtain the following commutative diagram

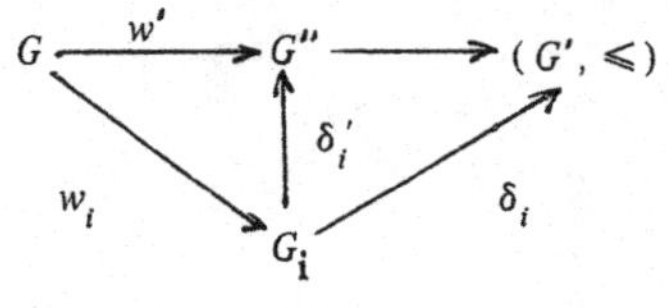

and δ_i is an o-homomorphism. We define a multivalued addition $\oplus'$ on G' :

$\oplus' =$ $\oplus'_m$ (see 3.9) iff for every $x,y,z \in G_1$, $a,b,c \in G_2$ such that $\delta_1(x) = \delta_1(y) = \delta_1(z)$, $\delta_2(a) = \delta_2(b) = \delta_2(c)$, we have $x \in y \oplus_{G_1} z$, $a \in b \oplus_{G_2} c$;

$\oplus' =$ $\oplus_m$, otherwise

Then $(G', \oplus')$ is a d-group and δ_i is a d-homomorphism, $(G', w) \in W(G)$, $(G', w) \leqslant (G_1, w_1), (G_2, w_2)$.

Let $(K, v) \in W(G)$ be such that $(K, v) \leqslant (G_1, w_1), (G_2, w_2)$, and let τ_1, τ_2 be d-homomorphisms such that the following diagram commutates.

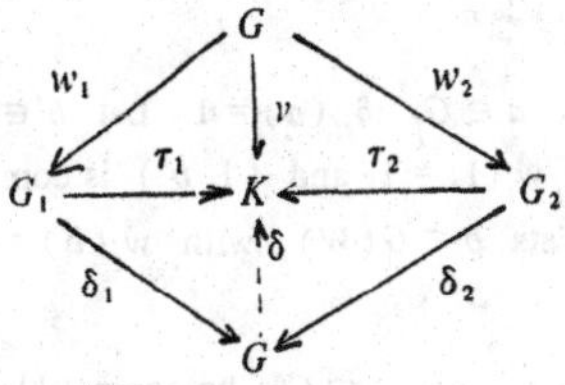

For $w(x) \in G'$ we set $\delta(w(x)) = v(x)$. It is easy to see that this definition of δ is correct and, moreover, δ is an o-homomorphism. We show in two cases that δ is o-homomorphism.

(1) $\oplus' = \oplus'_m$. If $\oplus_K = \oplus_m$, then clearly δ is a d-homomorphism. Let $\oplus_K = \oplus'_m$ and we assume that there exist $w(a) > w(b)$, $\delta(w(a)) = \delta(w(b))$. Then

$\tau_1 w_1(a) = v(a) = v(b) = \tau_2 w_2(b)$, $w_1(a) > w_1(b)$

and $w_1(b) \in w_1(a) \oplus_{G_1} w_1(b)$. Since τ_1 is a d-homomorphism, we have

$v(a) \in v(a) \oplus_K v(b) = v(a) \oplus'_m v(a)$,

a contradiction. Therefore, δ is a d-homomorphism in this case.

(2) $\oplus' = \oplus_m$. Then there exist $w_1(a), w_1(b), w_1(c) \in G_1$ such that $\delta_1 w_1(a) = \delta_1 w_1(b) = \delta_1 w_1(c)$, $w_1(a) \in w_1(b) \oplus_{G_1} w_1(c)$. Since $\delta\delta_1$ is a d-homomorphism, we obtain $\oplus_K = \oplus_m$ and δ is a d-valuation.

Therefore, $(K, v) \leqslant (G', w)$ and $(G', w) = (G_1, w_1) \wedge (G_2, w_2)$ in $W(G)$. ■

Now, let G be a d-group, $\alpha = (g_1, \ldots, g_n) \in G^n$. We say that α is *compatible with respect* to $((G_1, w_1), \ldots, (G_n, w_n)) \in W(G)^n$ if for every $1 \leqslant i, j \leqslant n$, $i \neq j$, $(G_{ij}, w_{ij}) = (G_i, w_i) \wedge (G_j, w_j) \in W(G)$ we have

$w_{ij}(g_i) = w_{ij}(g_j)$.

We observe that this definition is correct. For, if $(G', w') \leqslant (G_{ij}, w_{ij}) \leqslant (G', w')$ in $W(G)$ then from the fact $w_{ij}(g_i) = w_{ij}(g_j)$ it follows $w'(g_i) = w'(g_j)$.

In the following we denote by δ_{ij} the d-homomorphism such that $w_{ij} = \delta_{ij} \cdot w_i$. Further, for $W \subset W(G)$ we set

$G(W) = \{g \in G : w'(g) \geqslant 1 \text{ for every } (G', w') \in W\}$.

6.2. PROPOSITION. *Let G be a directed d-group and let $W \subset W(G)$. Then the following conditions are equivalent.*

(1) *For any* $\alpha = ((G_1, w_1), \ldots, (G_n, w_n)) \in W^n$, *and any* $(g_1, \ldots, g_n) \in G^n$ *compatible with respect to* α *and such that* $w_i(g_i) \geqslant 1$, $i = 1, \ldots, n$, *there exists* $g \in G(W)$ *such that* $w_i(g_i) = w_i(g)$, $i = 1, \ldots, n$.

(2) *For every* $(G_i, w_i), (G_j, w_j) \in W$, $a \in G_j$, $i \neq j$, *such that* $\delta_{ji}(a) = 1$ *there exists* $b \in G(W)$ *such that* $w_i(b) = 1$, $w_j(b) \geqslant a$.

Proof. (1) ⇒ (2). Let $(G_i, w_i), (G_j, w_j) \in W$, $a \in G_j$, $\delta_{ji}(a) = 1$. Let $b' \in B$ be such that $w_j(b') = a$. Then $\delta_{ji} w_j(b') = w_{ij}(b') = \delta_{ij} w_i(1) = 1$ and $(1, b')$ is compatible with respect to $((G_i, w_i), (G_j, w_j))$. Then there exists $b \in G(W)$ with $w_i(b) = w_i(1) = 1$, $w_j(b) = w_j(b') = a$.

(2) ⇒ (1). The proof is by induction on n. Let $(g_1, \ldots, g_n) \in G^n$ be compatible with respect to α, $w_i(g_i) \geqslant 1$. Then we may suppose $(G_k, w_k) \leqslant (G_j, w_j)$, for $k \neq j$. Further, if we suppose that there exists j, $2 \leqslant j \leqslant n$, with $\operatorname{Ker} \delta_{j1} = \{1\}$; from the fact that $(g_1, \ldots, g_j, \ldots, g_n)$ is a compatible with respect to $((G_1, w_1), \ldots, (G_{1j}, w'), \ldots, (G_n, w_n))$ where $(G_{1j}, w') = (G_1, w_1) \wedge (G_j, w_j)$, and by the induction, there exists $g \in G(W)$ such that $w_t(g) = w_t(g_t)$, $1 \leqslant t \leqslant n$, $t \neq j$. Since $(G_{1j}, w) \leqslant (G_j, w_j)$, we have $w'(g) = w'(g_j)$. Thus, $\delta_{j1} w_j(g) = w'(g) = w'(g_j) = \delta_{j1} w_j(g_j)$ and $w_j(g) = w_j(g_j)$.

Thus, we may suppose that for every $j \geqslant 2$ there exists $d_j \in G_j$, $d_j > 1$, $\delta_{j1}(d_j) = 1$. Then it is possible to show for every i, $1 \leqslant i \leqslant n$, the existence of $a_i \in G(W)$ with

$$w_i(a_i) = w_i(g_i) \;,$$

$$w_k(a_i) > w_k(g_n), \quad k \neq i .$$

Let

$$g \in a_1 \oplus \cdots \oplus a_n \subset G(W) .$$

Since w_i is a d-homomorphism, we have $w_i(g) \geqslant \min_k \{w_i(a_k)\} = w_i(a_i) = w_i(g_i)$. If $w_i(g) > w_i(g_i)$, we have $w_i(g_i) \in (\oplus_{k \neq i} w_i(a_k)) \oplus w_i(g)$, and since G_i is a local d-group, we obtain $w_i(g) \in \oplus_{k \neq i} w_i(a_k)$ and

$$w_i(g_i) \geqslant \min_{k \neq i} \{w_i(a_k)\} > w_i(a_i) = w_i(g_i) \text{, a contradiction.}$$ ∎

6.3. LEMMA. *Let* G *be a directed* d*-group and let* $W \subset W(G)$ *be such that* w *is an* o*-epimorphism and* $\operatorname{Ker} w$ *is a directed subgroup of* G *for every* $(G', w) \in W$. *Then the equivalent conditions of* 6.2 *are satisfied*.

Proof. Let $(G_i, w_i), (G_j, w_j) \in W$, $a \in G_j$, be such that $\delta_{ji}(a) = 1$. Let $(G_{ij}, w) = (G_i, w_i) \wedge (G_j, w_j)$. Since w_i and w_j are o-epimorphisms, the preorder relation on the factor group $G'' = G/(\operatorname{Ker} w_i \cdot \operatorname{Ker} w_j)$ defined in 6.1 is clearly an order relation and G_{ij} coincides with G''. Then there exist $b, a_i, a_j \in G$ such that $w_i(a_i) = 1$, $w_j(a_j) = 1$, $b = a_i \cdot a_j$, $w_j(b) = a$. Since $\operatorname{Ker} w_i$ is directed, there exists $c \in \operatorname{Ker} w_i$ such that $c \geqslant 1$, $c \geqslant a_i$. Then

$$w_i(c) = 1\,, \quad w_j(c) \geqslant w_j(a_i) = w_j(b a_j^{-1}) = w_j(b) = a\,. \qquad \blacksquare$$

6.4. **LEMMA.** *Let G be a directed d–group and let $(G_i, w_i) \in W(G)$, $i = 1, \ldots, n$, $a \in G$. Then there exists $b \in G$ such that*

$$w_i(b) = 1\,, \ \textit{for} \ \ w_i(a) \geqslant 1\,,$$
$$w_i(b) \leqslant w_i(a)\,, \ \textit{for} \ \ w_i(a) < 1\,.$$

Proof. Let $w_i(a) \geqslant 1$ for $1 \leqslant i \leqslant k$, $w_i(a) < 1$ for $k{+}1 \leqslant i \leqslant n$. For $n \in Z_+$ and $c \in G$ we set

$$nc = c \oplus c \oplus \cdots \oplus c \ \ (n \text{ times})\,,$$

and for $1 \leqslant i \leqslant k$ we set

$$P_i = \{ c \in G : w_i(c) > 1 \}\,.$$

Now, it there exist $n_{i,1}, \ldots, n_{i,t_i} \in Z_+$ such that for some $c_i \in G$ the following holds :

$$c_i \in (1 \oplus n_{i,1} \cdot a \oplus \cdots \oplus n_{i,t_i} \cdot a^{t_i} \oplus a^{t_i+1}) \cap P_i\,,$$

we denote β_i the first set in the intersection, otherwise, we set $\beta_i = \{1\}$. Further, let

$$b \in 1 \oplus a^2 c_1 \cdots c_k$$

where $c_j = 1$ for $\beta_j = \{1\}$. If $\beta_i = \{1\}$ for every i, then $b \in 1 \oplus a^2$. Hence, for $w_i(a) \geqslant 1$ we have $w_i(b) \geqslant 1$ and since $(1 \oplus a^2) \cap P_i = \phi$, we obtain $w_i(b) = 1$. If $w_i(a) < 1$, then from the fact $w_i(b) \in 1 \oplus_i w_i(a^2)$ it follows $w_i(b) = w_i(a^2) \leqslant w_i(a)$. Let $\beta_i \neq \{1\}$ for $i = 1, \ldots, p$; $1 \leqslant p \leqslant k$. We set

$$A = \bigoplus_{j = s_1 + \cdots + s_p} (n_{1,s_1} \cdots n_{p,s_p} a^j) = 1 \oplus n_1 a \oplus \cdots \oplus n_{t-1} a^{t-1} \oplus a^t\,,$$

where $t = t_1 + \cdots + t_p + p$, $n_{i,0} = n_{i,t_i+1} = 1$, $0 \leqslant s_i \leqslant t_i + 1$. Then

$$b \in 1 \oplus a^2 c_1 \cdots c_p \subset 1 \oplus a^2 \beta_1 \cdots \beta_p \subset 1 \oplus a^2 A\,.$$

Let $1 \leqslant i \leqslant k$. Then $w_i(a) \geqslant 1$. If $\beta_i = \{1\}$, we have $(1 \oplus a^2 A) \cap P_i \neq \phi$ and $w_i(b) = 1$. If $\beta_i \neq \{1\}$, we have $w_i(a^2 c_1 \cdots c_p) \geqslant w_i(c_i) > 1$ and it follows $w_i(b) = 1$. Let $k{+}1 \leqslant i \leqslant n$. Since

$$b \in 1 \oplus a^2 \oplus n_1 a^3 \oplus \cdots \oplus n_{t-1} a^{t+1} \oplus a^{t+2}$$

there exist $x_1 \in n_1 a^3, \ldots, x_{t-1} \in n_{t-1} a^{t+1}$ such that

$$b \in 1 \oplus a^2 \oplus x_1 \oplus \cdots \oplus x_{t-1} \oplus a^{t+2}\,.$$

Since $w_i(x_j) \in n_j \cdot w_i(a^{j+2})$; $j = 1, \ldots, t-1$, $w_i(a^{j+2}) > w_i(a^{t+2})$,

we have

$$w_i(x_j) > w_i(a^{t+2})\ ;\quad j = 1, \dots, t-1\,,$$

and

$$w_i(a^{t+2}) < w_i(x_j),\ 1,\ w_i(a^2)\ ;\ j = 1, \dots, t-1\,.$$

Therefore, $w_i(b) = w_i(a^{t+2}) < w_i(a)$. ∎

6.5. PROPOSITION. *Let* $(G, \oplus)$ *be a directed d-group,* $(G_i, w_i) \in W(G)$, $i = 1, \dots, n$ $g_i \in w_i^{-1}(G_i^+)$. *Then the following conditions are equivalent.*

(1) *There exists* $g \in G$ *such that* $w_i(g) = w_i(g_i)$, $i = 1, \dots, n$.

(2) $(g_1, \dots, g_n)$ *is compatible with respect to* $((G_1, w_1), \dots, (G_n, w_n))$.

Proof. (2) ⇒ (1). Let $G'_+ = \bigcap_{i=1}^{n} w_i^{-1}(G_i^+)$. Then $(G'_+, \oplus)$ is an m-ring of G. Let $G' = D(G'_+)$. Hence, $G' = \{gU : g \in G\}$ where $U = \{h \in G : w_i(h) = 1\ ;\ i = 1, \dots, n\}$. For every i we define

$$w'_i : G' \longrightarrow G_i\ ;\ w'_i(gU) = w_i(g)\,.$$

It is easy to see that $(G_i, w'_i) \in W(G')$. Further, $\mathrm{Ker}\, w'_i$ is directed. For, by 6.4, for every $gU \in \mathrm{Ker}\, w'_i$ there exists $h \in G$ such that $w_j(h) = 1$ for $w_j(g) \geqslant 1$ and $w_k(h) \leqslant w_k(g)$ for $w_k(g) < 1$. Then $hU \in \mathrm{Ker}\, w'_i$ and $hU \leqslant U$, gU in G'. A d-valuation w'_i is an o-epimorphism. In fact, for $gU \in G'$ such that $w'_i(gU) = w_i(g) \geqslant 1$ there exists $b \in G$ (by 6.4) such that

$$w_k(b) = 1 \quad \text{for} \quad w_k(g) \geqslant 1\,,$$

$$w_j(b) \leqslant w_j(g) \quad \text{for} \quad w_j(g) < 1\,.$$

We set $h = gb^{-1}$. If $w_k(g) \geqslant 1$, then $w_k(h) = w_k(g)$, and for $w_j(g) < 1$ we have $w_j(h) \geqslant 1$. Thus $hU \geqslant U$ in G' and $w'_i(hU) = w_i(h) = w'_i(gU)$.

We set

$$W = \{(G_1, w'_1), \dots, (G_n, w'_n)\} \subset W(G')\,.$$

Let $(g_1, \dots, g_n) \in G^n$ be compatible with respect to $((G_1, w_1), \dots, (G_n, w_n))$ and let for $1 \leqslant i, j \leqslant n$, $i \neq j$ $(G_{ij}, w) = (G_i, w_i) \wedge (G_j, w_j)$ in $W(G)$. Then for the canonical d-valuation w' from G' onto G_{ij} we have

$$(G_{ij}, w') = (G_i, w'_i) \wedge (G_j, w'_j)$$

in $W(G')$. Since $w(g_i) = w(g_j)$ implies $w'(g_i U) = w'(g_j U)$, we obtain the proposition by 6.3, and 6.2.

(1) ⇒ (2). Trivial. ∎

6.6. COROLLARY. *Let* G *be a directed d-group,* $(G_i, w_i) \in W(G)$, $g_i \in G$; $i = 1, \dots, n$. *Then the following conditions are equivalent.*

(1) *There exists* $g \in G$ *such that* $w_i(g) = w_i(g_i)$; $i = 1, \dots, n$.

(2) $(g_1, \dots, g_n)$ *is compatible with respect to* $((G_1, w_1), \dots, (G_n, w_n))$.

P r o o f. (2) ⇒ (1). We may assume that for some k, $1 \leqslant k \leqslant n$; the following holds:

$w_i(g_i) \geqslant 1$; $i = 1, \dots, k$,

$w_j(g_j) < 1$; $j = k+1, \dots, n$.

Then for $(G_{tj}, w_{tj}) = (G_t, w_t) \wedge (G_j, w_j)$, $t \neq j$, we have

$1 \leqslant w_{i,k+1}(g_i) = w_{i,k+1}(g_{k+1}) \leqslant 1$; $i = 1, \dots, k$,

$1 \geqslant w_{k,j}(g_j) = w_{k,j}(g_k) \geqslant 1$; $j = k+1, \dots, n$,

and it follows that $(g_1, \dots, g_k, 1, \dots, 1)$, $(1, \dots, 1, g_{k+1}^{-1}, \dots, g_n^{-1})$ are compatible with respect to $((G_1, w_1), \dots, (G_n, w_n))$. According to 6.5, there exist $a,b \in G$ such that

$w_i(a) = w_i(g_i)$; $i = 1, \dots, k$,

$w_j(a) = 1$; $j = k+1, \dots, n$,

$w_i(b) = 1$; $i = 1, \dots, k$,

$w_j(k) = w_j(g_j^{-1})$; $j = k+1, \dots, n$.

Then $w_i(ab^{-1}) = w_i(g_i)$; $i = 1, \dots, n$.

(1) ⇒ (2). Trivial. ∎

6.7. LEMMA. *Let* G *be a directed d-group,* $H, H_1, \dots, H_n \in \mathrm{M}(G)$, $H_1^+ \cap \cdots \cap H_n^+ \subset H^+$. *Then there exists* i, $1 \leqslant i \leqslant n$, *such that* $H_i \subset H$.

P r o o f. The proof is by induction on n. Let $n = 2$, $H_1^+ \cap H_2^+ \subset H^+$ and suppose that $H_i^+ \not\subset H^+$, i.e. there exist $x \in H_1^+ \setminus H$, $y \in H_2^+ \setminus H$. Since H is prime, $(x \oplus y) \cap H = \phi$ by 4.2, and $(x \oplus y) \cap H_1^+ \cap H_2^+ = \phi$. Let $z \in x \oplus y$. Then $x \in (z \oplus y) \cap H_1$, $y \notin H_1$, and we have $z \in H_1^+$. Analogously, $z \in H_2^+$, a contradiction.

Let $n \geqslant 3$. If for some i, $1 \leqslant i \leqslant n$, $\bigcap_{j \neq i} (H^+ \cup H_j^+) \subset H^+ \cup H_i^+$ holds, then $\bigcap_{j \neq i} H_j^+ \subset H^+$, and the induction hypothesis implies $H_k \subset H$ for some k. Now, suppose that for every i, $1 \leqslant i \leqslant n$, there exists $z_i \in \bigcap_{j \neq i} (H^+ \cup H_j^+) \setminus (H^+ \cup H_i^+)$.

Let $z \in z_1 \oplus z_2 \cdots z_n$. Since H is convex and prime, we obtain $z \notin H$. Hence, there exists i such that $z \notin H_i$. Again, since H_i is convex and prime, it is easy to see that this is a contradiction. The conclusion of 6.7 follows from the case considered previously and by the induction. ∎

6.8. PROPOSITION. *Let* G *be a directed d-group,* $(G_i, w_i) \in W(G)$; $i = 1, \dots, n$, $A = \bigcap_{i=1}^{n} w_i^{-1}(G_i^+)$. *Then* $D(A)$ *is a Prüfer d-group.*

P r o o f. Let $G' = D(A)$. It is easy to see that

$$P_i = \{gU \in G' \ : \ w_i(g) > 1\}; \ i = 1, \ldots, n,$$

are prime ideals of G'_+. Let $H_i = \psi^{-1}(P_i) \in \mathrm{M}(G')$ and let $w'_i \ : \ G' \longrightarrow G_i$ be defined by $w'_i(gU) = w_i(g)$. Then w'_i is clearly an d-homomorphism. Since $\mathrm{Ker}\, w'_i = H_i$ is a convex subgroup, w'_i is an o-epimorphism and $G'/H \cong_o G_i$ is an o-group. Let $H \in \mathrm{M}(G')$. Then

$$\{1\} = \bigcap_{i=1}^{n} H_i^+ \subset H^+$$

and $H_i \subset H$ for some i by 6.7. In this case G'/H is an o-group and G' is a Prüfer d-group by 5.11. ∎

6.9. REMARKS. (1) As W. Krull [80] conjectured, for any l-group G and prime l-ideals $H_1, \ldots, H_n$ of G for a family $(g_1H_1, \ldots, g_nH_n) \in G/H_1 \times \ldots \times G/H_n$ such that

$$g_iH_iH_j = g_jH_iH_j \quad ; \ i, j = 1, \ldots, n, \ i \neq j,$$

there exists an element $g \in G$ such that

$$gH_i = g_iH_i \quad ; \ i = 1, \ldots, n.$$

The first proof of this conjecture was done by D. Müller [96]. We observe that a proof of this conjecture may be done using 6.6. In fact, we denote by w_i the canonical d-homomorphism $G \longrightarrow G/H_i = G_i$. Then $((G_i, \oplus_m), w_i) \in W(G, \oplus_m)$ by 3.14.(2). According 6.6, there exists $g \in G$ such that $w_i(g) = w_i(g_i)$.

(2) Let G be an l-group, $\{H_i : i \in J\}$ be a set of prime l-ideals of G such that $\bigcap_{i \in J} H_i = \{1\}$. Let $i_1, \ldots, i_n \in J$, $(g_1, \ldots, g_n) \in G^n$ be such that $g_kH_{i_k}H_{i_t} = g_tH_{i_k}H_{i_t}$, $g_kH_{i_k} \geqslant H_{i_k}$; $k, t = 1, \ldots, n$, $k \neq t$. Analogously as given above for $G_r = G/H_{i_r}$, $((G_r, \oplus_m), w_r) \in W(G, \oplus_m)$. Since $\mathrm{Ker}\, w_r = H_{i_r}$ is directed, and w_r is an o-epimorphism, it follows by 6.2, 6.3, that there exists $g \in G_+ \doteq G(\{(G_r, w_r) \ : \ r \in J\})$ such that $gH_{i_k} = g_kH_{i_k}$; $k = 1, \ldots, n$, and we obtain another type of the approximation theorem for l-groups.

(3) Again, let G be an l-group. $H_1, \ldots, H_n$ be prime l-ideals of G. Then $H_i \in \mathrm{M}(G, \oplus_m)$ by 4.9.(1). We consider a d-group $D(A)$ as in 6.8 where $w_i: G \longrightarrow G/H_i$ is a canonical l-homomorphism. It is easy to see that $D(A)$ is an l-group but we do not know whether the multivalued addition $\oplus$ of $D(A)$ coincides with $\oplus_m$ It is clear that this is equivalent to the following:

For every $x, y, z \in G$, $a, b \in \cap H_i$ $(1 \leqslant i \leqslant n)$ such that

$$x \wedge y = a(x \wedge z) = b(y \wedge z),$$

there exist $c, d \in \cap H_i$ $(1 \leqslant i \leqslant n)$ such that

$cx \wedge dy = z \wedge cx = z \wedge dy$.

(4) P. Ribenboim [115] has published one of the first approximation theorem for valuations of the field K:

For valuations $w_1, \ldots, w_n$ of K and $(\alpha_1, \ldots, \alpha_n) \in G(R_{w_1}) \times \ldots \times G(R_{w_n})$ the following is equivalent.

(a) There exists $x \in K$ such that $w_i(x) = \alpha_i$, $1 \leqslant i \leqslant n$.

(b) $(\alpha_1, \ldots, \alpha_n)$ is a compatible family (for definition see Ribenboim [115],p. 7).

It is possible to create the proof using 6.6. In fact, for $A = \cap R_{w_i}$ $(1 \leqslant i \leqslant n)$, the canonical map

$$w_i : (G(A), \oplus_A) \longrightarrow (G(R_{w_i}), \oplus_m) = G_i$$

$$w_i(w_A(x)) = w_i(x),\ x \in K,$$

is clearly a d-homomorphism and if $\alpha_i = w_i(a_i)$, $a_i \in K$, it follows (in the proof (b) $\Rightarrow$ $\Rightarrow$ (a)) that $(w_A(a_1), \ldots, w_A(a_n))$ is compatible with respect to $((G_1, w_1), \ldots$ $\ldots, (G_n, w_n))$. According to 6 6. there exists $w_A(x) \in G(A)$ such that

$$w_i(x) = w_i(w_A(x)) = w_i(w_A(a_i)) = w_i(a_i) = \alpha_i,\ 1 \leqslant i \leqslant n.$$

(5) In the theory of commutative domains it is well known that for valuations $w_1, \ldots, w_n$ of a field K, a ring $A = \cap R_{w_i}$ $(1 \leqslant i \leqslant n)$ is a Prüfer domain (see Gilmer [42], 18.8). The proof may be done using 6.8.

In fact, let $G = (G(A), \oplus_A)$ and let w_i be the same as in (4). Then

$$\bigcap_{i=1}^{n} w_i^{-1}(G(R_{w_i})_+) = G_+$$

$$D(G_+) = (G, \oplus_A),$$

and $(G, \oplus_A)$ is a Prüfer d-group by 6.8. According to 5.12.1, A is a Prüfer domain.

As we have observed in 6.9, in a commutative algebra and in a theory of l-groups there are several approximation theorems which have almost the same structure. When we compare all these approximation theorems we can point out that there exists a common background for these theorems. Roughly speaking, all these theorems (and many other too) have the following structure:

There is a family of elements of a form $(\alpha_i)_{i \in I}$, $\alpha_i \in B_i$, which are in some sense compatible with respect to B_i and the approximation theorem is then a map which assigns to every such compatible family a solution from some structure B. It is interesting that in all cases mentioned above these compatibility conditions may be derived using pushouts (or pullbacks under some modification) in a suitable category. Moreover, in this case these compatibility conditions become formally the same, hence it would be natural to define a general approximation theorem using this pushout-like definition. Unfortunately, there are approximation theorems (for example the generalized Ribenboim approximation theorem [115],Theoréme 5) which may not be expressed using this definition. So we need a more general definition,and one of possible way is to cancel the compatibility conditions.

On the other hand, although we cancel the compatibility condition from the definition of approximation theorem, in many cases it remains conserved and this enables us to create a notion of a representation of an approximation theorem by a sheaf (or presheaf) over some site and a localization. This representation is in many cases rather faithful since it enables us to translate a relation between two associated sheaves to a relation between original approximation theorems.

Moreover, since every presheaf is a model of a many-sorted language, we may describe some relations between approximation theorems using the local properties of corresponding models.

At first, we deal with an abstract form of approximation theorems which will include several known types of approximation theorems for valuations, l-groups and d-groups. As we have observed, an approximation theorem is a map, which assigns to a compatible family $(\alpha_i)_{i\in I}$ of elements of some kinds a solution of this family, i.e.. an element x which equals under some maps f_i, $i \in I$, elements α_i. For our further investigation we need some categorical modification of this description.

We call a category J *quasidiscrete* provided that $|J| = J \cup \{*\}$, where J is a set, $* \in J$, and the only nontrivial morphisms in J are $i \longrightarrow *$, $i \in J$. Then we have the following definition.

6.10. DEFINITION. Let J be a quasidiscrete category and let A, B be functors:

$$A, B : J^{op} \longrightarrow \mathrm{SET}\,.\quad A(i) = B(i),\ i \in J.$$

Then a natural transformation

$$\xi : A \longrightarrow B$$

such that $\xi_i = 1_{A(i)}$, $i \in J$, and ξ_* is a surjection is called an *approximation theorem* (or A T).

Roughly speaking, if $\xi : A \longrightarrow B$ is an approximation theorem, the set $A(*)$ (always denoted by A) is assumed to be the set of compatible families from sets $A(i)$ (always denoted by A_i ($= B_i = B(i)$)) for $i \in J$, the set $B(*)$ ($= B$) is assumed to be the set of solutions for these compatible families and $\xi_* : A \longrightarrow B$ assigns a compatible family to its solution. Hence, the following diagram commutates.

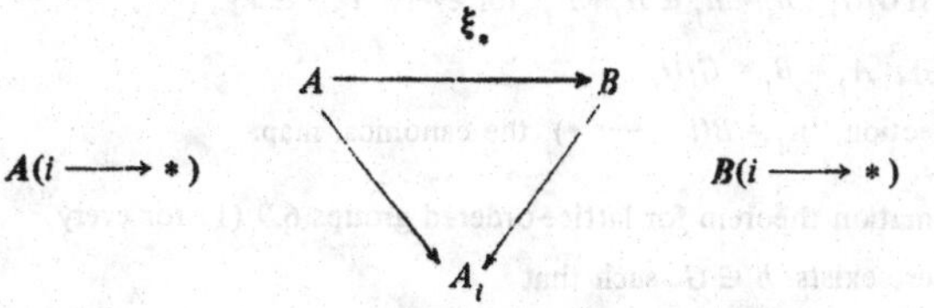

The following couple of examples shows that many of classical approximation theorems are A T in our sense too.

6.11. **EXAMPLE.** Let n be a natural number, $J = \{1, \ldots, n\}$ and let w_i be a valuation of a field $K(i \in J)$ with a valuation ring R_i and a value group G_i. We use the notation from 6.9 (4). Let

$A = \{(w_i(a_i))_{i \in J} \in \Pi G_i\ (i \in J) : (w_A(a_1), \ldots, w_A(a_n))$ is compatible with respect to $((G_1, w_1), \ldots, (G_n, w_n))\}$,

$A_i = B_i = G_i$, $B = G(A)$,

$A(i \longrightarrow *) : A \longrightarrow G_i$ be the ith projection,

$B(i \longrightarrow *) = w_i$.

Then using the Ribenboim's approximation theorem, 6.9.(4), for every $\alpha = (\alpha_i)_{i \in I} \in A$ there exists $x \in K$ such that $w_i(x) = \alpha_i$ $(i \in J)$. Hence, $w_A(x) \in B$ and we may set

$\xi(\alpha) = w_A(x)$.

Clearly, $\xi : A \longrightarrow B$ is an A T

6.12. **EXAMPLE.** Let J and w_i be the same as in 6.11. We set

$A = \{(\alpha_i)_{i \in J} \in \Pi G_i^+ : \delta_{ij}(\alpha_i) = \delta_{ji}(\alpha_j)$ for every $i, j \in J\}$,

$B = G(R)_+$, $A_i = B_i = G_i^+$,

$A(i \longrightarrow *) =$ projection, $B(i \longrightarrow *) = w_i$.

Again, according to 6.5, for every $\alpha = (\alpha_i)_{i\in J} \in A$ there exists $x \in R$ such that $w_i(x) = \alpha_i$, $i \in J$. Thus, for $\xi_*(\alpha) = w_R(x) \in B$ we have an A T.

6.13. **EXAMPLE.** Let $J = \{1, \ldots, n\}$ and let G be an l-group, H_i, $i \in J$, prime l-ideals of G. We set

$$A = \{(h_i + H_i)_{i\in J} \in \Pi\, G/H_i : h_i - h_j \in H_i + H_j \text{ for every } i, j \in J\},$$
$$B = G/(\cap H_i\ (i \in J)),\quad A_i = B_i = G/H_i,$$
$$A(i \longrightarrow *) = \text{projection},\quad w_i = B(i \longrightarrow *) \text{ the canonical map.}$$

Then using Krull approximation theorem for lattice-ordered groups 6.9 (1), for every $\alpha = (h_i + H_i)_{i\in J} \in A$ there exists $h \in G$ such that

$$h_i + H_i = h + H_i,\quad i = 1, \ldots, n.$$

Then $\xi_*(\alpha) = h + (\cap H_i) \in B$ and

$$w_i(\xi_*(\alpha)) = h_i + H_i,\quad i = 1, \ldots, n.$$

Hence, we obtain an A T.

6.14. **EXAMPLE.** Let $J = \{1, \ldots, n\}$ and let G be a d-group. Let $(G_1, w_1), \ldots, (G_n, w_n) \in W(G)$, $R = \cap w_i^{-1}(G_i^+)$. Then R is a m-ring in G. We set

$$A = \{(a_1, \ldots, a_n) \in G^n : (a_1, \ldots, a_n) \text{ is compatible with respect to } ((G_1, w_1), \ldots, (G_n, w_n))\},$$
$$B = D(R),\quad A_i = B_i = G_i,$$
$$A(i \longrightarrow *) = w_i \cdot \text{projection}_i,\quad B(i \longrightarrow *) = w_i, \text{ where}$$
$$w_i(w_R(x)) = w_i(x).$$

Then by approximation theorem 6.5. for every $\alpha = (a_1, \ldots, a_n) \in A$ there exists $x \in G$ such that $w_i(x) = w_i(a_i)$, $i = 1, \ldots, n$. Hence, for $\xi_*(\alpha) = w_R(x) \in B$ we have

$$w_i(\xi_*(\alpha)) = w_i(x) = w_i(a_i) = w_i \cdot \text{projection}_i(\alpha) \text{ and we obtain an A T.}$$

6.15. **EXAMPLE.** Let J and w_i be the same as in 6.11. We set $A_i = B_i = G_i$, $i \in J$. On the set $\Pi\, G_i (i \in J) \times K^n$ we define an equivalence relation $\equiv$ such that

$$(a_1, \ldots, a_n, b_1, \ldots, b_n) \equiv (a_1, \ldots, a_n, b_1, \ldots, b_n) \text{ iff } a_i = a_i$$

$i \in J$, and if there exists $x \in K$ such that $w_i(x - b_i) = a_i$ for every $i \in J$ then there exists a $x \in K$ such that $w_i(x - b_i) = a_i$ for every $i \in J$ and conversely, and, moreover, $w_i(b_i - b_i) \geqq a_i$ for every $i \in J$.

Elements of $\Pi G_i (i \in J) \times K^n / \equiv$ we denote by $[a_1, \ldots, a_n, b_1, \ldots, b_n]$. Then we set

$$A = \{[a_1, \ldots, a_n, b_1, \ldots, b_n] : w_i \wedge w_j(b_i - b_j) \geqq \delta_{ij}(a_j) \text{ for every } i, j \in J\}.$$

It may be easily done that this definition is correct. Moreover, on the set K^{n+1} we define an equivalence relation $\sim$ such that

$$(x, b_1, \ldots, b_n) \sim (x', b'_1, \ldots, b'_n) \text{ iff } w_i(b_i - b'_i) \geqq w_i(x - b_i) = w_i(x' - b'_i)$$
for every $i \in J$.

Elements of $K^{n+1}/\sim$ we denote by $[x, b_1, \ldots, b_n]$. Then we set

$$B = \{[x, b_1, \ldots, b_n] : \delta_{ij} w_i(x - b_i) = \delta_{ji} w_j(x - b_j) \text{ for every } i, j \in J\}.$$

Analogously, it may be observed that this definition is correct. Further, we set

$$A(i \longrightarrow *)([a_1, \ldots, a_n, b_1, \ldots, b_n]) = a_i,$$
$$B(i \longrightarrow *)([x, b_1, \ldots, b_n]) = w_i(x - b_i).$$

Then according to Ribenboim's approximation theorem, Ribenboim [115]; Theoréme 5, for every $\alpha = [a_1, \ldots, a_n, b_1, \ldots, b_n] \in A$ there exists $x \in K$ such that $w_i(x - b_i) = a_i$, $i \in J$. We set

$$\xi_*(\alpha) = [x, b_1, \ldots, b_n] \in B.$$

Then $\xi : A \longrightarrow B$ is a natural transformation and ξ_* is a surjection. In fact, let $[x, b_1, \ldots, b_n] \in B$. Then according to Ribenboim [115], Theoréme 5, there exists $y \in K$ such that $w_i(y) = w_i(x - b_i)$ for every $i \in J$. Hence, $\beta = [w_i(y), \ldots, w_n(y), b_1, \ldots, b_n] \in A$ and $\xi_*(\beta) = [x, b_1, \ldots, b_n]$.

Therefore, ξ is an A T.

Further we deal with a relation between approximation theorems and sheaves (or presheaves). This relation is in many cases rather close since it enables us to translate some relations between two approximation theorems in some relations between corresponding presheaves. Moreover, since every presheaf is a model of some canonical many-sorted language we may describe some relations between A T using logical properties of corresponding models.

At first, we recall some basic facts about sheaves. Let $\mathbf{C}$ be a *site*, i.e. $\mathbf{C}$ is a small-complete category with a Grothendieck topology and let F be a *presheaf* over $\mathbf{C}$, i.e. F is a contravariant functor from $\mathbf{C}$ to the category SET of all small sets. Let $\mathbf{a} = (s_i \xrightarrow{f_i} s)_{i \in J} \in \mathrm{Cov}(s)$ be a localization of an element $s \in /\mathbf{C}/$. Then a system of natural transformations $\alpha = (\alpha_i)_{i \in J}$; $\alpha_i : \mathrm{Hom}\ (-, s_i) \longrightarrow F$, is called a *compatible system from* $\mathbf{a}$ *to* F if for every $i, j \in J$ the diagram

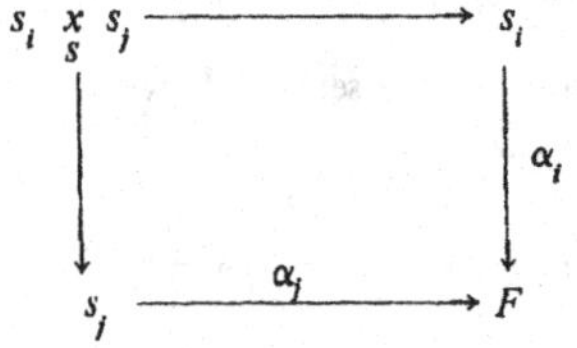

commutates, where $s_i \underset{s}{\times} s_j$ is a pullback of s_i, s_j over s. (We identify hom-functor Hom $(-, x)$ with an object x.) Then a natural transformation $x : s \longrightarrow F$ is a *solution* of α provided that for every $i \in J$ we have $x \,.\, f_i = \alpha_i$. Using the formalism of Yoneda embedding, every natural transformation $y : s \longrightarrow F$ $(s \in /\mathbf{C}/)$ may be identify with an element of $F(s)$, namely with $y_s(1_s)$. We frequently use this identification.

Finally, recall that a presheaf F is a *sheaf* provided that for every localization $\mathbf{a}$ and every compatible system α from $\mathbf{a}$ to F there exists the unique solution of α.

Let $\mathbf{C}$ be a site, $\mathbf{a} = (s_i \xrightarrow{f_i} s)_{i \in J} \in \mathrm{Cov}(s)$ be a localization. Then $\mathbf{a}$ may be considered as a functor

$$\mathbf{a} : J^{op} \longrightarrow \mathbf{C}^{op}$$

such that $\mathbf{a}(i) = s_i$, $\mathbf{a}(*) = s$, $\mathbf{a}(i \longrightarrow *) = f_i$. With this notation we have the following definition of a representation of an A T.

6.16. DEFINITION. Let $\xi : A \dashrightarrow B$ be an A T over a quasidiscrete category J. We say that ξ is *represented by a presheaf* $F : \mathbf{S}^{op} \longrightarrow \mathrm{SET}$ *over a site* $\mathbf{S}$ *and a localization* $\mathbf{s} = (s_i \longrightarrow s)_{i \in J} \in \mathrm{Cov}(s)$ if there exist natural transformations r, ρ such that r is epi and the diagram

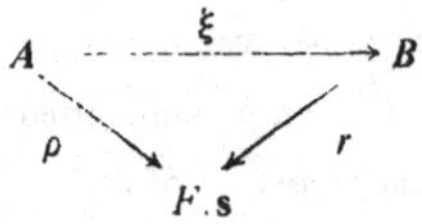

commutates.

We observe at first that every sheaf creates an A T. which is represented by this sheaf. Moreover, a presheaf F is called a *quasisheaf* if for every localization $\mathbf{a}$ and every compatible system α from $\mathbf{a}$ to F, there exists at least one solution of α. In this case we have the following proposition.

6.17. PROPOSITION. *Let F be a quasisheaf over* $\mathbf{S}$. *Then for every localization* $\mathbf{s} = (s_i \xrightarrow{f_i} s)_{i\in J}$ *there exists an A T over J represented by F over* $\mathbf{s}$.

P r o o f. On the set $F(s)$ we define an equivalence relation $\sim$ such that

$x \sim y$ iff x and y are the common solution of a compatible system from $\mathbf{s}$ to F.

Then we set

$A = \{\alpha = (\alpha_i)_{i\in J} \in \Pi F(s_i) : \alpha$ is a compatible system from $\mathbf{s}$ to $F\}$,
$A_i = B_i = F(s_i)$, $B = F(s)/\sim$,
$A(i \longrightarrow *) =$ projection$_i$, $B(i \longrightarrow *) = f_i$, where
$f_i(\tilde{x}) = F(f_i)(x)$ and $x \in \tilde{x} \in B$.

Further we set

$\xi_*((\alpha_i)_i) = \tilde{x}$, where x is a solution of $(\alpha_i)_i$,
$r_* : F(s) \longrightarrow F(s)/\sim$ is the canonical map,
$\rho_* : A \longrightarrow F(s)$, $\alpha \longrightarrow$ solution of α.

Then $r : F\cdot \mathbf{s} \longrightarrow B$, $\rho : A \longrightarrow F\cdot\mathbf{s}$ are natural transformations and we have $r \cdot \rho = \xi$. Hence, ξ is an A T which is represented by F over $\mathbf{s}$. ∎

For the following theorem we need some notation. Let $B : J^{op} \longrightarrow \mathrm{SET}$ be a functor. Let there exists a concrete category $\mathbf{K}$ such that B_i is an underlying set for some object in $\mathbf{K}$ for every $i \in J$ and let in $\mathbf{K}$ pushouts exist for every pair of morphisms $g_i = B(i \longrightarrow *)$, $g_j = B(j \longrightarrow *)$, $i, j \in J$. Let

$$\begin{array}{ccc} B & \xrightarrow{\ g_i\ } & B_i \\ {\scriptstyle g_j}\downarrow & & \downarrow{\scriptstyle \delta_{ij}} \\ B_j & \xrightarrow{\ \delta_{ji}\ } & B_{ij} \end{array}$$

be a pushout diagram for every $i, j \in J$ and let T with morphisms $t_i : T \longrightarrow B_i$, $i \in J$, be a pullback of morphisms δ_{ij} in the category SET, i.e.

(a) for every $i, j \in J$, $\delta_{ij} \cdot t_i = \delta_{ji} \cdot t_j$,

(b) if $X \in /\mathrm{SET}/$ with morphisms $h_i : X \longrightarrow B_i$ such that $\delta_{ij} \cdot h_i = \delta_{ji} \cdot h_j$ for every $i, j \in J$ then there exists the unique $t : X \longrightarrow T$ such that $t_i \cdot t = h_i$, $i \in J$.

Then we may consider a functor

$$(F =) \ \mathrm{pull}_{\mathrm{SET}} (\mathrm{push}_{\mathbf{K}} B) : J^{op} \longrightarrow \mathrm{SET}$$

such that $F(i) = B_i$, $F(*) = T$, $F(i \longrightarrow *) = t_i$.

Further, let $\mathbf{K}_1$, $\mathbf{K}_2$, $\mathbf{K}_3$ be categories and let F, F_1, G, G_1 be functors,

$$\mathbf{K}_1 \underset{G}{\overset{F}{\rightrightarrows}}\ \alpha \ \mathbf{K}_2 \underset{G_1}{\overset{F_1}{\rightrightarrows}}\ \beta \ \mathbf{K}_3$$

$\alpha : F \longrightarrow G$, $\beta : F_1 \longrightarrow G_1$ natural transformations. Recall that a *horizontal composition* of α and β, in symbol $\beta \circ \alpha$, is a natural transformation

$$\beta \circ \alpha : F_1 F \longrightarrow G_1 G$$

such that

$$(\beta \circ \alpha)_x = G_1(\alpha_x) \cdot \beta_{F(x)}, \quad x \in /\mathbf{K}_1/ .$$

Finally, we need some possibility of comparing various approximation theorems and, namely, we need to specify the notion that an approximation theorem may be derived from an other approximation theorem . The following definition gives an exact description.

6.18. DEFINITION. Let $\xi : A \longrightarrow B$ ($\xi' : A' \longrightarrow B'$) be an approximation theorem over I (J). Then ξ *may be derived from* ξ' if there exist a functor $\pi : J^{op} \longrightarrow I^{op}$ and natural transformations u, ν such that the diagram

$$\begin{array}{ccc} A \,.\, \pi & \xrightarrow{\ \xi \circ 1_\pi\ } & B \,.\, \pi \\ {\scriptstyle u}\downarrow & & \uparrow{\scriptstyle \nu} \\ A & \xrightarrow{\ \xi'\ } & B \end{array}$$

commutates.

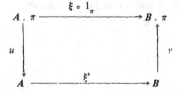

Roughly speaking, when we want to find a solution of a compatible system from A in ξ, we may create a new compatible system using u, then find its solution in ξ' and from this solution create a solution of an original compatible system using v.

Then we have the following theorem.

6.19. THEOREM. *Let* $\xi : A \longrightarrow B$ *be an approximation theorem and let there exists a natural transformation*

$$r : \text{pull}_{\text{SET}}(\text{push}_{\mathbf{K}} B) \longrightarrow B .$$

Then there exists an approximation theorem $\bar{\xi}$ *which may be derived from* ξ *and such that it is represented by a sheaf.*

Proof. Let $T = \text{pull}(\text{push}\, B)$. Then for every $i, j \in J$ we have $\delta_{ij} \cdot g_i = = \delta_{ji} \cdot g_j$ and there exists a morphism (in SET) $u : B \longrightarrow T(*)$ such that $pr_i \cdot u = = g_i$. (Here $T(*)$ may be identify with a set $\{(\alpha_i)_{i \in J} \in \Pi B_i : \delta_{ij}(\alpha_i) = \delta_{ji}(\alpha_j)$ for every $i, j \in J\}$ and $t_i = pr_i$.) For $a, b \in B$ we define

$$a \sim b \text{ iff } u(a) = u(b)$$

and let

$$C : J^{op} \longrightarrow \text{SET}$$

be such that

$$C(i) = B_i ,\ C(*) = B/\sim ,\ C(i \longrightarrow *) = \bar{g}_i ,\ \text{where}$$
$$\bar{g}_i(\bar{x}) = g_i(x) \quad (x \in \bar{x} \in C(*)) .$$

Let

$$\bar{\xi}_* : A \longrightarrow B/\sim$$
$$\bar{\xi}_*(a) = \xi_*(a) \in B/\sim ,\ \bar{\xi}_i = 1 .$$

Then $\bar{\xi} : A \longrightarrow C$ is a natural transformation and if $v : B \longrightarrow C$ is such that $v_i = 1$, $v_* =$ canonical map, than

$$v \cdot \xi \cdot 1_A = \bar{\xi} \text{ and } \bar{\xi} \text{ may be derived from } \xi .$$

Now, let $\mathbf{R}_0$ be the set of all finite subsets of J. For $K \in \mathbf{R}_0$ we set

$$B_K = \text{pushout}_{\mathbf{K}}\{B(i \longrightarrow *) : i \in K\} ,$$

and let $\delta_{i,K}$, $\delta_{j,K}$ be morphisms in $\mathbf{K}$ such that

$$\begin{array}{ccc} B & \xrightarrow{\quad g_i \quad} & B_i \\ {\scriptstyle g_j}\downarrow & & \downarrow{\scriptstyle \delta_{i,K}} \\ B_j & \xrightarrow[\quad]{\delta_{j,k}} & B_K \end{array}$$

is a pushout diagram. We set $g_K = \delta_{i,K} \cdot g_i$, where $i \in K$. Let $\mathbf{R}$ be the set of all isomorphism classes of $\mathbf{R}_0$ with an element $* \notin \mathbf{R}_0$. We define a preorder relation $\leqq$ on $\mathbf{R}$ such that

$K \leqq K'$ iff there exists (in $\mathbf{K}$) a morphism $\delta_{K',K}$ such that $\delta_{K',K} \cdot g_{K'} = g_K$,
$K \leqq *$ for every $K \in \mathbf{R}$.

With respect to this relation $\leqq$ the set $(\mathbf{R}, \leqq)$ is a complete category with

$\text{pullback}_{\mathbf{R}} \{ i : i \in K \} = K$,

where $i = \{i\}$. Moreover, we set

$\text{Cov}_0(K) = \{(K' \longrightarrow K)_{K \in \mathbf{R}'} : \mathbf{R}' \subseteq \mathbf{R} \ , \ K \in \mathbf{R}'\}$,
$\text{Cov}_0(*) = \{(i \longrightarrow *)_{i \in J}\}$.

Then $\mathbf{P} = \{ \text{Cov}_0(K) : K \in \mathbf{R} \}$ generates a Grothendieck topology on $\mathbf{R}$ and the system $\mathbf{P}$ is stable under pullback. Hence, $\mathbf{R}$ is a site. We define a functor

$F : \mathbf{R}^{op} \longrightarrow \text{SET}$

such that

$F(K) = B_K$, $F(*) = T(*)$, $F(K' \longrightarrow K) = \delta_{K,K'}$,
$F(K \longrightarrow *) = \delta_{i,K} \cdot \text{pr}_i$ (where $i \in K$) .

Then F is a sheaf. In fact, it is necessary to deal with a localization $\mathbf{i} = (i \longrightarrow *)_{i \in J}$ only and in this case we have

$F(*) = \{$compatible family from $\mathbf{i}$ to $F\}$.

Now, for $a \in A$ we set $\rho_*(a) = (f_i(a))_{i \in J} \in F(*)$, $\rho_i = 1$. Hence, $\rho : A \longrightarrow F$ is a natural transformation. Then the diagram

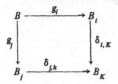

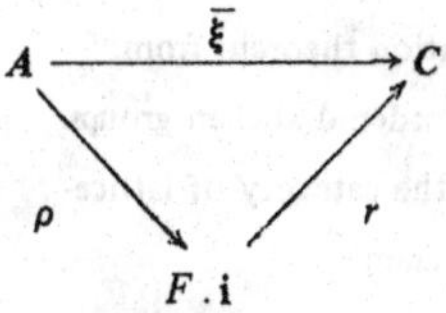

commutates, where $r_*(a) = \overline{r_*(a)} \in B/\sim$. In fact, for $a \in A$ we have

$$\bar{g}_i r_* \cdot \rho_*(a) = \bar{g}_i \cdot r_*((f_i(a))_i) = \mathrm{pr}_i((f_i(a))_i) = f_i(a) = g_i \,.\, \xi_*(a) = \bar{g}_i \,.\, \bar{\xi}_*(a)$$

for every $i \in J$. Hence, we have $\mathrm{pr}_i \cdot \mu_*(r_* \rho_*(a)) = \mathrm{pr}_i \mu_*(\xi_*(a))$ for every $i \in J$ and it follows $\mu_*(r_* \rho_*(a)) = \mu_*(\xi_*(a))$, $\bar{\xi}_*(a) = \bar{r}_* \rho_*(a)$. Therefore, $\bar{\xi}$ is represented by F. ∎

6.20. COROLLARY. *Let* $\xi : A \longrightarrow B$ *be an approximation theorem such that*

$$\mathrm{pull}_{\mathrm{SET}}(\mathrm{push}\ B) \cong B\,.$$

Then ξ *is represented by a sheaf.*

Proof. In fact, since B is a pullback of δ_{ij}, $i, j \in J$, then the map μ from the proof of 6.19 is a bijection, i.e. $a \sim b \Leftrightarrow a = b$. Hence, $C = B$. ∎

6.21. PROPOSITION. *Approximation theorems from 6.11 – 6.14 are represented by sheaves and for approximation theorem* ξ *from 6.15 there exists an A T* $\bar{\xi}$ *which may be derived from* ξ *and such that it is represented by a sheaf.*

Proof. In fact, for ξ from 6.11, for an element $w_R(x) \in G(R) = B$ we set

$$r_*(w_R(x)) = (w_i(x))_{i \in J} \in \mathrm{pull}_{\mathrm{SET}}(\mathrm{push}\ B) \ (*)$$
$$r_i = 1\,.$$

Then using an approximation theorem from 6.11 it may be observed that r_* is a bijection and, therefore,

$$r : \mathrm{pull}_{\mathrm{SET}}(\mathrm{push}_{\mathbf{K}} B) \cong B$$

is a natural equivalence, where $\mathbf{K}$ is the category of totally ordered abelian groups with ordered homomorphisms. Hence, ξ is represented by a sheaf according to 6.20.

Analogously the proof may be done in the case of approximation theorem from 6.12, where $\mathbf{K}$ will be the category of positive cones of totally ordered abelian groups.

In the case of approximation theorem from 6.13 let $\mathbf{K}$ be the category of lattice-ordered groups with lattice-ordered homomorphisms. Then the diagram

$$\begin{array}{ccc} G/H_0 & \longrightarrow & G/H_i \\ \big\downarrow & & \big\downarrow \\ G/H_j & \longrightarrow & G/(H_i + H_j) \end{array}$$

is a pushout diagram for every $i, j \in J$, $H_0 = \cap H_i \ (i \in J)$ and the map

$$\begin{aligned} r_* : G/H_0 &\longrightarrow \{(h_i + H_i)_i \in \Pi\, G/H_i : h_i - h_j \in H_i + H_j\} \\ g + H_0 &\longmapsto (g + H_i)_{i \in J} \end{aligned}$$

is a bijection. Hence,

$$\text{pull}_{\mathbf{SET}}(\text{push } B) \cong B$$

and ξ is represented by a sheaf.

Let $\mathbf{K} = (W(G), \leqq)$ in case of approximation theorem from 6.14. Then the rest of the proof is almost analogical to the previous one.

Finally, let us consider the approximation theorem ξ from 6.15. Then for $\mathbf{K} = \mathbf{SET}$ the diagram

$$\begin{array}{ccc} B & \xrightarrow{\ g_i\ } & G_i \\ {\scriptstyle g_j}\big\downarrow & & \big\downarrow{\scriptstyle \delta_{ij}} \\ G_j & \xrightarrow{\ \delta_{ji}\ } & G_{ij} = G_{w_i \wedge w_j} \end{array}$$

is a pushout for every $i, j \in J$. In fact, since for every $[x, b_1, \ldots, b_n] \in B$, $\delta_{ij} w_i(x - b_i) = \delta_{ji} w_j(x - b_j)$ holds for every $i, j \in J$, the above diagram commutates. Let $X \in /\mathrm{SET}/$ with maps t_i, t_j be such that $t_i \cdot g_i = t_j \cdot g_j$. Let $R = \cap R_i \ (i \in J)$ and let $G(R)$ be the group of divisibility of R. For $w_R(x) \in G(R)$ we set

$$s(w_R(x)) = [x, 0, \ldots, 0] \in B.$$

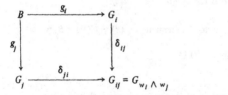

Then this definition is correct and we have

$$g_i \cdot s = w_i\,, \quad i \in J\,,$$

where $w_i(w_R(x)) = w_i(x)$. Hence, we have the following commutative diagram.

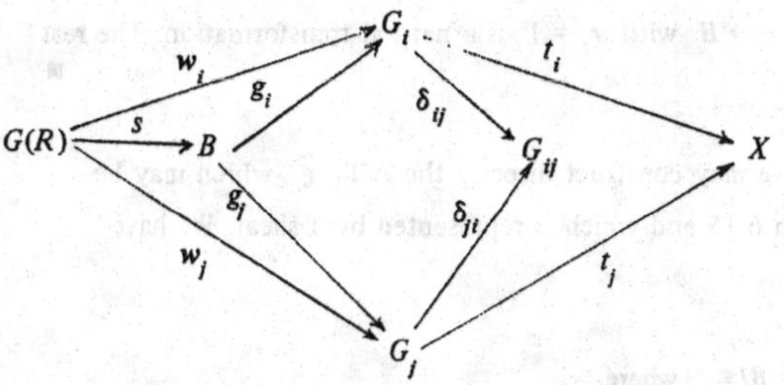

Hence, $t_i \cdot w_i = t_j \cdot w_j$. Since R is a Prüfer ring, there are convex directed subgroups H_i, H_j of $G(R)$ such that $G_i = G(R)/H_i$, $G_j = G(R)/H_j$, $G_{ij} = G(R)/(H_i + H_j)$ and w_i, w_j are canonical maps. Now, for $\alpha + (H_i + H_j) \in G_{ij}$ we set

$$t(\alpha + (H_i + H_j)) = t_i(\alpha + H_i)(= t_i w_i(\alpha) = t_j w_j(\alpha) = t_j(\alpha + H_j)) \in X\,.$$

This definition is correct. In fact, let $\alpha + (H_i + H_j) = \beta + (H_i + H_j)$. Then there exist $h \in H_i$, $k \in H_j$ such that $\beta = \alpha + h + k$ and we obtain

$$t_i(\beta + H_i) = t_i(\alpha + h + k + H_i) = t_i(\alpha + k + H_i) = t_j(\alpha + \ k + H_j) =$$
$$= t_j(\alpha + H_j) = t_i(\alpha + H_i)\,.$$

Then $\delta_{ij} t = t_i$, $\delta_{ji} t = t_j$ and t is evidently the unique map with these properties. Hence, G_{ij} is a pushout in SET.

Now, let $T = \text{pull}_{\text{SET}}(\text{push}_{\text{SET}}\, B)$ and we may assume that

$$T(*) = \{(\alpha_i)_{i \in J} \in \Pi\, G_i : \delta_{ij}(\alpha_i) = \delta_{ji}(\alpha_j)\,, \quad i, j \in J\}\,,$$
$$T(i) = G_i\,, \quad T(i \longrightarrow *) = \text{pr}_i\,.$$

For $(\alpha_i)_i \in T(*)$ we set

$$r_*((\alpha_i)_i) = [a, 0, \ldots, 0] \in B\,,$$

where $a \in K$ is such that $w_i(a) = \alpha_i$, $i \in J$. Then commutates the diagram

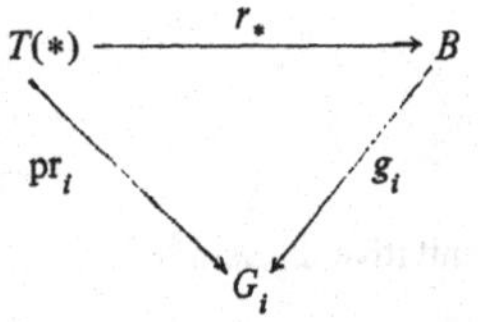

and $r : \text{pull}_{\text{SET}} (\text{push}_{\text{SET}} B) \longrightarrow B$ with $r_i = 1$ is a natural transformation. The rest follows from 6.19. ∎

Using the proof of 6.19 we may construct directly the A T $\bar{\xi}$ which may be derived from the A T ξ from 6.15 and which is represented by a sheaf. We have

$$\bar{\xi} : A \longrightarrow C,$$

where $C(i) = B_i = G_i$, $C(*) = B/\equiv$, where

$$[x, b_1, \ldots, b_n] \equiv [x', b'_1, \ldots, b'_n] \text{ iff } w_i(x - b_i) = w_i(x' - b'_i) \quad i \in J.$$

Then $B/\equiv\, = G(R)$. In fact, let elements of $B/\equiv$ be denoted by $\{x, b_1, \ldots, b_n\}_0$, where $[x, b_1, \ldots, b_n] \in B$. Then a map

$$B/\equiv \longrightarrow G(R)$$
$$\{x, b_1, \ldots, b_n\}_0 \longmapsto w_R(y),$$

where $y \in K$ is such that $w_i(x - b_i) = w_i(y)$, $i \in J$, is a bijection, and the approximation theorem ξ is such that diagram

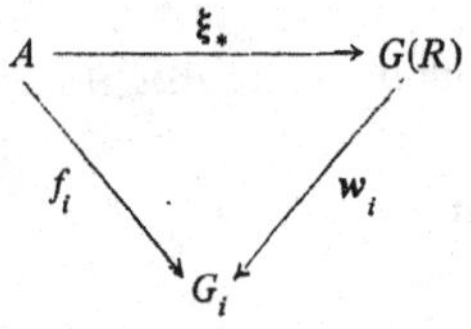

commutates, i.e. for $[a_1, \ldots, a_n, b_1, \ldots, b_n] = \alpha \in A$ we have

$$\xi_*(\alpha) = w_R(a),$$

where $a \in K$ is such that $w_i(a) = w_i(a_i)$, $i \in J$.

Let ξ and ξ' be A T which are represented by presheaves G and F, respectively. In the next theorem we show sufficient conditions dealing with G

and F under which ξ may be derived from ξ'.

At first, we need some notation.

Let $F, G:\ \mathsf{E}^{op} \longrightarrow \mathrm{SET}$ be functors, $r : F \longrightarrow G$ a natural transformation. Then in the category $\mathrm{SET}^{\mathsf{E}^{op}}$ there exists a pullback pull(r, r) of the diagram

$$\begin{array}{ccc} F & \xrightarrow{\ \ r\ \ } & G \\ {\scriptstyle p}\uparrow & & \uparrow{\scriptstyle r} \\ \mathrm{pull}(r,r) & \xrightarrow{\ \ p\ \ } & F \end{array}$$

and this functor pull(r, r) is a subfunctor of $F^2 :\ \mathsf{E}^{op} \longrightarrow \mathrm{SET}$ with

$$F^2(j) = F(j) \times F(j),\quad F^2(j \longrightarrow j') = F(j \longrightarrow j') \times F(j \longrightarrow j'),$$

since for $u : j \longrightarrow j'$ in E we have

$$\mathrm{pull}(r, r)(j) = \{(x, y) \in F(j) \times F(j) : r_j(x) = r_j(y)\},$$
$$\mathrm{pull}(r, r)(u)(x, y) = (F(u)(x),\ F(u)(y)).$$

Hence, for various natural transformations r from F we may functors pull(r, r) compare as subobjects of F^2.

Further, let $\mathsf{C}_1, \mathsf{C}_2$ be sites, $S :\ \mathsf{C}_1 \longrightarrow \mathsf{C}_2$ be a functor. We say that S *creates localizations* provided that for every $a \in |\mathsf{C}_2|$, $\alpha = (a_i \xrightarrow{f_i} a)_{i \in I} \in$ $\in \mathrm{Cov}(a)$ there exist $b \in |\mathsf{C}_1|$, $\beta = (b_j \xrightarrow{g_j} b)_{j \in J} \in \mathrm{Cov}(b)$ and a map $\pi : J \longrightarrow I$ such that

(i) $S(b_j) = a_{\pi(j)}$, $j \in J$

(ii) $S(b) = a$,

(iii) $S(g_j) = f_{\pi(j)}$, $j \in J$.

In this case we write $S(\beta) = \alpha$.

Now, let

$$\xi : A \longrightarrow B,\ A, B : I^{op} \longrightarrow \mathrm{SET}$$
$$\xi' : A' \longrightarrow B',\ A', B' : J^{op} \longrightarrow \mathrm{SET}$$

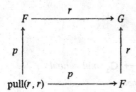

be approximation theorems which are represented by presheaves G and F, respectively, over the sites $\mathsf{C}_1, \mathsf{C}_2$ and localizations $\mathbf{a}$, $\mathbf{b}$, respectively. Let r, ρ, r', ρ' be the natural transformations such that the diagrams

$$\begin{array}{ccc} A & \xrightarrow{\ \xi\ } & B \\ & {\scriptstyle\rho}\searrow \quad \nearrow{\scriptstyle r} & \\ & G\,.\,\mathrm{a} & \end{array} \qquad\qquad \begin{array}{ccc} A' & \xrightarrow{\ \xi'\ } & B' \\ & {\scriptstyle\rho'}\searrow \quad \nearrow{\scriptstyle r'} & \\ & F\,.\,\mathrm{b} & \end{array}$$

commutates. Then the following theorem holds.

6.22. THEOREM. *Let there exist a functor* $S : \mathsf{C}_1 \longrightarrow \mathsf{C}_2$ *and natural transformations* $\alpha : G\,.\,S \longrightarrow F$, $\beta : F \longrightarrow G\,.\,S$ *such that*

(1) $S(\mathrm{b}) = \mathrm{a}$,

(2) $\beta\,.\,\alpha = 1$,

(3) $\mathrm{pull}(r', r') \lneqq \mathrm{pull}(r\,.\,\beta, r\,.\,\beta)$.

Then ξ *may be derived from* ξ'.

Proof. Let $\mathrm{a} = (a_i \xrightarrow{f_i} a)_{i \in I}$, $\mathrm{b} = (b_j \xrightarrow{t_j} b)_{j \in J}$ and let $\pi : J \twoheadrightarrow I$ be such that satisfies (i) - (iii). We define a natural transformation $u : A \cdot \pi \longrightarrow A'$ in the following way.

For $j \in J$ we set

$$u_j = r'_j \cdot \alpha_{b_j} \cdot \rho_{\pi(j)} : B_{\pi(j)} \longrightarrow B'_j .$$

Further, for $a \in A$ there exists $a' \in A'$ such that

$$r'_\bullet \cdot \alpha_b \cdot \rho_\bullet(a) = \xi'_\bullet(a') \qquad (1)$$

and using some choise of this a' we may define a map

$$u_\bullet : A \longrightarrow A', \quad a \longmapsto a' .$$

Then the diagram

$$\begin{array}{ccc} A & \xrightarrow{\ u_\bullet\ } & A' \\ {\scriptstyle f_{\pi(j)}}\downarrow & & \downarrow{\scriptstyle f'_j} \\ B_{\pi(j)} & \xrightarrow{\ u_j\ } & B'_j \end{array}$$

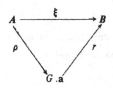

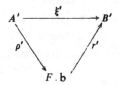

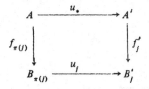

commutates, since we have

$$u_j f_{\pi(j)} = r'_j \alpha_{b_j} \rho_{\pi(j)} f_{\pi(j)} = r'_j \alpha_{b_j} G(h_{\pi(j)}) \rho_* = r'_j F(t_j) \alpha_b \rho_* = g'_j r'_* \alpha_b \rho_* =$$
$$= g'_j \xi'_* u_* = f'_j u_* \,.$$

Hence, u is a natural transformation.

Further, we define $\nu : B' \longrightarrow B \cdot \pi$ such that

$$\nu_j = r_{\pi(j)} \beta_{b_j} \rho'_j \,, \quad j \in J \,.$$

All elements of B' are of the form $r'_*(x')$ for some $x' \in F(b)$. Then we put

$$\nu_*(r'_*(x')) = r_* \beta_b(x') \,.$$

This definition is correct. In fact, let $r'_*(x') = r'_*(y')$, then $(x', y') \in \text{pull}(r', r')\ (*) \subseteq$ $\subseteq \text{pull}(r\beta, r\beta)\ (*)$ and it follows $r_* \beta_b(x') = r_* \beta_b(y')$. Then the diagram

$$\begin{array}{ccc}
B'_j & \xrightarrow{\ \nu_j\ } & B_{\pi(j)} \\
\uparrow{\scriptstyle g'_j} & & \uparrow{\scriptstyle g_{\pi(j)}} \\
B' & \xrightarrow{\ \nu_*\ } & B
\end{array}$$

commutates. In fact, for $r'_*(x') \in B'$ we have

$$\nu_j g'_j r'_*(x') = r_{\pi(j)} \beta_{b_j} \rho'_j g'_j r'_*(x') = r_{\pi(j)} \beta_{b_j} \rho'_j r'_j F(t_j)\,(x') =$$
$$= r_{\pi(j)} \beta_{b_j} F(t_j)\,(x') = r_{\pi(j)} G(h_{\pi(j)}) \beta_b(x') =$$
$$= g_{\pi(j)} r_* \beta_b(x') = g_{\pi(j)} \nu_* r'_*(x') \,.$$

Hence, ν is a natural transformation. Finally, we have the following commutative diagram.

$$\begin{array}{ccc}
A \cdot \pi & \xrightarrow{\ \xi \circ 1\ } & B \cdot \pi \\
\downarrow{\scriptstyle u} & & \uparrow{\scriptstyle \nu} \\
A' & \xrightarrow{\ \xi'\ } & B'
\end{array}$$

In fact, for $j \in J$ we have

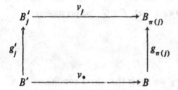

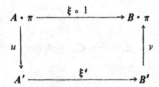

$$(\xi \circ 1)_j = \xi_{\pi(j)} = 1$$

$$(\nu \cdot \xi' \cdot u)_j = \nu_j \cdot u_j = r_{\pi(j)} \beta_{b_j} \rho'_j r'_j \alpha_{b_j} \rho_{\pi(j)} = 1 \ ,$$

and for $a \in A$ we have

$$\nu_* \xi'_* u_*(a) = \nu_* r'_*(\alpha_b \, \rho_*(a)) = r_* \, \beta_b(\alpha_b \, \rho_*(a)) = r_* \, \rho_*(a) = \xi_*(a) = (\xi \circ 1)_*(a) \ .$$

Therefore, ξ may be derived from ξ'. ■

6.23. COROLLARY. *Let* $\mathbf{C}_1 = \mathbf{C}_2$, $\mathbf{a}=\mathbf{b}$ *and let* $\alpha : G \longrightarrow F$, $\beta : F \longrightarrow G$ *be such that*

(1) $\beta \cdot \alpha = 1$,

(2) $\mathrm{pull}(r\,,\,r) \lessgtr \mathrm{pull}(r\beta,\,r\beta)$.

Then ξ *may be derived from* ξ'.

The natural question arising here is the following. Let F, G be presheaves. Under which conditions does the implication

F sheaf $\Rightarrow$ G sheaf

hold? In the following theorem and in its corollary we show that these conditions are analogical to those of 6.22 and 6.23.

6.24. THEOREM. *Let* F (G) *be a quasisheaf (presheaf) over* $\mathbf{C}_1$ $(\mathbf{C})$. *Let the following conditions hold.*

(1) There exists a functor $S : \mathbf{C}_1 \longrightarrow \mathbf{C}$ *such that*

(a) S *is continuous,*

(b) S *creates localizations,*

(2) There exists a natural transformation

$$\varphi : \mathrm{Hom}_{\mathbf{C}_1}(-,-) \longrightarrow \mathrm{Hom}_{\mathbf{C}}(-,-) \, . \, (S \times S)$$

such that for each $b \in /\mathbf{C}_1/$ *there exists a functor* $P_b : \mathbf{C}_1 \longrightarrow \mathbf{C}_1$ *such that (a)* $\varphi_{b,b}(1_b) = 1_{S(b)}$,

(b) $S \cdot P_b = S$,

(c) $\varphi_{-,b} \circ 1_{P_b} : b \cdot P_b \longrightarrow S(b) \cdot S$ *is a natural equivalence,*

(3) There exist natural transformations $\alpha : G \cdot S \longrightarrow F$, $\beta : F \longrightarrow G \cdot S$ *such that* $\beta \cdot \alpha = 1$.

Then

(i) G *is a quasisheaf,*

(ii) *If* F *is a sheaf then* G *is a sheaf.*

Proof. (i) Let $\mathbf{p} = (a_i \xrightarrow{f_i} a)_{i \in I} \in \mathrm{Cov}(a)$, $a \in /\mathbf{C}/$, and let $(a_i \xrightarrow{\xi_i} G)_{i \in I}$ be a compatible family from $\mathbf{p}$ to G. Since S creates localizations, there exists $\mathbf{e} = (b_j \xrightarrow{g_j} b)_{j \in J} \in \mathrm{Cov}(b)$ and a map π with the properties (i)-(iii). For $j \in J$ we set

$$\eta_j = \alpha \cdot (\xi_{\pi(j)} \circ 1_S) \cdot \varphi_{-,b_j} .$$

Hence, $\eta_j : b_j \longrightarrow F$ is a natural transformation and, moreover, $(b_j \xrightarrow{\eta_j} F)_{j \in J}$ is a compatible system. In fact, let for $j, k \in J$, the diagram

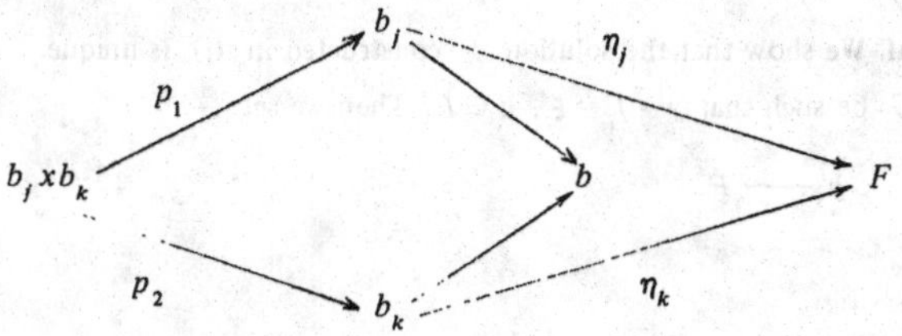

be a pullback. Since S is continuous, $S(b_j x b_k)$ is a pullback of $a_{\pi(j)}$, $a_{\pi(k)}$. Then we have

$$\begin{aligned} \eta_j p_1 &= \alpha \cdot (\xi_{\pi(j)} \circ 1_S) \cdot \varphi_{-,b_j} \cdot p_1 = \alpha \cdot ((\xi_{\pi(j)} \cdot S(p_1) \circ 1_S) \cdot \varphi_{-,b_j x b_k} = \\ &= \alpha \cdot ((\xi_{\pi(k)} \cdot S(p_2) \circ 1_S) \cdot \varphi_{-,b_j x b_k} = \\ &= \alpha \cdot (\xi_{\pi(k)} \circ 1_S) \cdot \varphi_{-,b_k} \cdot p_2 = \eta_k \cdot p_2 . \end{aligned}$$

Let $x' \longrightarrow F$ be a solution of $(\eta_j)_{j \in J}$, i.e.

$$x' \cdot g_j = \eta_j, \ j \in J .$$

Let $x : a = S(b) \longrightarrow G$ be such that

$$x \circ 1_S = (\beta \circ 1_{P_b}) \cdot (x' \circ 1_{P_b}) \cdot (\varphi_{-,b} \circ 1_{P_b})^{-1} .$$

Since S as object function is a surjection, this relation determines x uniquely. Then

$$x \cdot f_i = \xi_i , \quad i \in I .$$

In fact, it suffices to show that

$$(x \cdot f_i) \circ 1_S = \xi_i \circ 1_S , \quad i \in I .$$

Using 'interchange law' we obtain for $i = \pi(j)$

$$\begin{aligned}(x \cdot f_i) \circ 1_S &= (x \circ 1_S) \cdot (f_{\pi(j)} \circ 1_S) = (\beta \circ 1_{P_b}) \cdot (x \circ 1_{P_b}) \cdot (\varphi_{-,b} \circ 1_{P_b})^{-1} \cdot \\ &(f_{\pi(j)} \circ 1_S) = (\beta \circ 1_{P_j}) \cdot (x \circ 1_{P_j}) \cdot (g_j \circ 1_{P_j}) \cdot (\varphi_{-,b} \circ 1_{P_j})^{-1} = \\ &= (\beta \circ 1_{P_j}) \cdot (x \cdot g_j \circ 1_{P_j}) \cdot (\varphi_{-,b} \circ 1_{P_j})^{-1} = (\beta \circ 1_{P_j}) \cdot (\eta_j \circ 1_{P_j}) \cdot (\varphi_{-,b} \circ 1_{P_j})^{-1} = \\ &= (\beta \circ 1_{P_j}) \cdot (\alpha \circ 1_{P_j}) \cdot (\xi_{\pi(j)} \circ 1_S) \cdot (\varphi_{-,b_j} \circ 1_{P_j}) \cdot (\varphi_{-,b_j} \circ 1_{P_j})^{-1} = \\ &= (\beta \cdot \alpha \circ 1_{P_j}) \cdot (\xi_{\pi(j)} \circ 1_S) = \xi_{\pi(j)} \circ 1_S , \quad \text{where } P_j = P_{b_j} .\end{aligned}$$

Hence, G is a quasisheaf.

(ii) Let F be a sheaf. We show that the solution x constructed in (i) is unique. In fact, let $y : Sb \longrightarrow G$ be such that $y \cdot f_i = \xi_i$, $i \in I$. Then we set

$$y' = \alpha \cdot (y \circ 1_S) \cdot \varphi_{-,b} : b \longrightarrow F .$$

Then

$$\begin{aligned}y' \cdot g_j &= \alpha \cdot (y \circ 1_S) \cdot \varphi_{-,b} \cdot g_j = \alpha \cdot ((y \cdot f_{\pi(j)} \circ 1_S) \cdot \varphi_{-,b_j} = \\ &= \alpha \cdot (\xi_{\pi(j)} \circ 1_S) \cdot \varphi_{-,b_j} = \eta_j , \quad j \in J\end{aligned}$$

and we have $x' = y'$. Thus,

$$y' \circ 1_{P_b} = (\alpha \circ 1_{P_b}) \cdot (y \circ 1_S) \cdot (\varphi_{-,b} \circ 1_{P_b})$$

and it follows

$$\begin{aligned}y \circ 1_S &= (\beta \circ 1_{P_b}) \cdot (y' \circ 1_{P_b}) \cdot (\varphi_{-,b} \circ 1_{P_b})^{-1} = \\ &= (\beta \circ 1_{P_b}) \cdot (x' \circ 1_{P_b}) \cdot (\varphi_{-,b} \circ 1_{P_b})^{-1} = x \circ 1_S .\end{aligned}$$

Therefore, $x = y$ and G is a sheaf. ∎

6.25. COROLLARY. *Let* F (G) *be a quasisheaf (presheaf) over* C *and let there exist natural transformations* $\alpha : G \longrightarrow F$, $\beta : F \longrightarrow G$ *such that* $\beta \cdot \alpha = 1$. *Then*

(i) G *is a quasisheaf,*

(ii) *If* F *is a sheaf then* G *is a sheaf.*

As it may be expected, the approximation theorem ξ_2 (= ξ from 6.12) may be derived from ξ_1 (= ξ from 6.11). Let F_1 (F_2) be a sheaf which represents ξ_1 (ξ_2) over the site R (here $\mathsf{R}_1 \cong \mathsf{R}_2$), i.e.

$$F_1 : \begin{array}{ccc} i & \longmapsto & G_i \\ \downarrow & & \downarrow w_i \\ * & \longmapsto & G(R) \end{array} \qquad\qquad F_2 : \begin{array}{ccc} i & \longmapsto & G_i^+ \\ \downarrow & & \downarrow w_i \\ * & \longmapsto & G(R)_+ \end{array}$$

We set $\alpha : F_2 \longrightarrow F_1$ such that α_i, α_* are natural injections and $\beta : F_1 \longrightarrow F_2$ such that

$\beta_i(a) = /a/$ (the absolute value)

$\beta_*(a) = b$,

where $b \in G(R)_+$ is such that $w_i(b) = /w_i(a)/$, $i = 1, \ldots, n$. Then $\beta \cdot \alpha = 1$ and clearly pull$(1, 1) \leqq$ pull(β, β). Hence, ξ_2 may be derived from ξ_1.

In the following we show a nontrivial example of 6.23. It is well known that a solution of ξ_1 may be derived using the solutions of ξ_2 twice. In fact, for every $(a_1, \ldots, a_n) \in A_1$ (= A from 6.11) it may be easily seen that $\mathrm{c} = (a_1 \vee 0, \ldots, a_n \vee 0)$, $\mathrm{d} = (-a_1 \vee 0, \ldots, -a_n \vee 0) \in A_2$ (= A from 6.12) and we have

$$\xi_2(\mathrm{c}) - \xi_2(\mathrm{d}) = \xi_1((a_1, \ldots, a_n)) .$$

The natural question arising here is whether the approximation theorem ξ_1 *may be derived from* ξ_2 (may be for another set of valuations $w_1, \ldots, w_n$) in our sense. The answer is in the affirmative and in what follows we construct such a family of valuations.

Hence, let $\xi_1 : A_1 \longrightarrow B_1$ be the approximation theorem from 6.11. For every w_i $(1 \leqq i \leqq n)$ we construct a valuation w_i' of a field $K(X)$ with the value group $Z \times G_i$ ordered lexicographically, where Z is the ordered group of integers, in such a way that for element of $K[X]$ we set

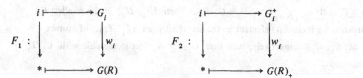

$$w_i'\Big(\sum_{j=0}^{m} a_j x^j\Big) = \min_j \{(j\,,\, w_i(a_j)) : a_j \neq 0\} \in Z \times G_i$$

Then according to Bourbaki [13], Ch. VI, § 10, Prop. 1, w_i' may be extended onto a valuation on a field $K(X)$ and if G_i is considered to be an ordered subgroup of $Z \times G_i$ ($a \longmapsto (0, a)$), w_i' is an extension of w_i. Now, let w' be a valuation constructed in such a way using a valuation w of K. Then we obtain

6.26. LEMMA. *For every valuations* w_1, w_2 *of* K *we have*

$$(w_1 \wedge w_2)' = w_1' \wedge w_2' .$$

Proof. Let $G_{w_1 \wedge w_2} = G_1/H_1 = G_2/H_2$. Then (H_1, H_2) is the smallest (in pairwise ordering by inclusion) element in the set of all pairs (T_1, T_2), of convex subgroups of G_1, G_2, respectively, such that G_1/T_1 is order isomorphic with G_2/T_2 and the diagram

$$\begin{array}{ccc} K^{\times} & \xrightarrow{\;w_1\;} & G_1 \\ {\scriptstyle w_2}\big\downarrow & & \big\downarrow{\scriptstyle \text{nat}} \\ G_2 & \xrightarrow{\;\text{nat}\;} & G_1/T_1 \cong_o G_2/T_2 \end{array}$$

commutates. We set

$$H_i' = \{0\} \times H_i \subset Z \times G_i\,, \quad i = 1, 2.$$

Then it is easily seen that

$$(Z \times G_1)/H_1' \cong Z \times (G_1/H_1) \cong Z \times (G_2/H_2) \cong (Z \times G_2)/H_2' .$$

Let w be the composition of the following maps.

$$K(X)^{\times} \xrightarrow{\;w_1'\;} Z \times G_1 \xrightarrow{\;\text{nat}\;} (Z \times G_1)/H_1'$$

Then w is a valuation of $K(X)$ and $w \leqq w_1', w_2'$, $w = (w_1 \wedge w_2)'$. Let $v = w_1' \wedge w_2'$. Then there exist convex directed subgroups Δ_i of $Z \times G_i$ and ordered isomorphism ω such that the diagram

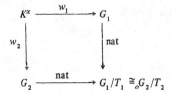

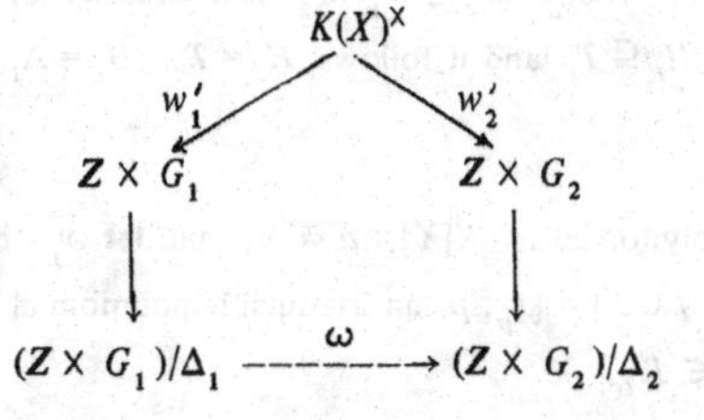

commutates. Since $(w_1 \wedge w_2)' \leqq w_1', w_2'$, we have $v \geqq (w_1 \wedge w_2)' = w$ and we have a commutative diagram

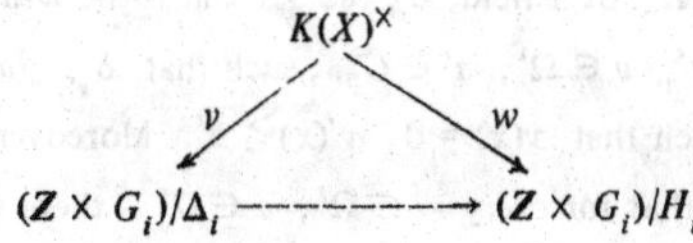

Hence, $\Delta_i \subseteq H_i'$ and it follows $\Delta_i = \{0\} \times T_i$ for some $T_i \subseteq H_i$. Then the following diagram commutates, where $\rho((0, w_1(a) + T_1)) = (0, w_2(a) + T_2)$.

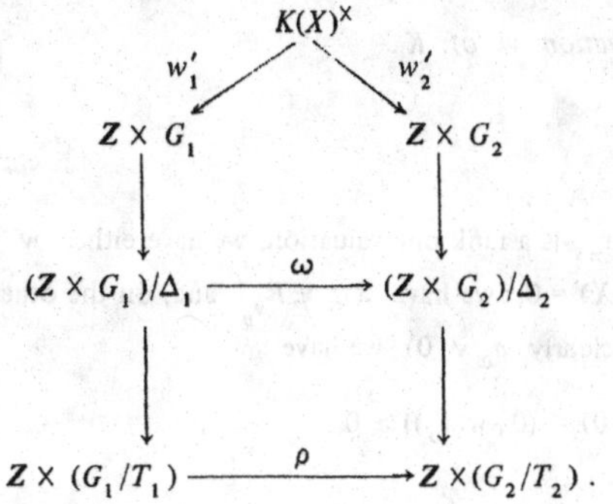

We define a map $\pi : G_1/T_1 \longrightarrow G_2/T_2$ such that $\pi(w_1(a) + T_1) = w_2(a) + T_2$. Then it is possible to show that π is an order isomorphism and the diagram

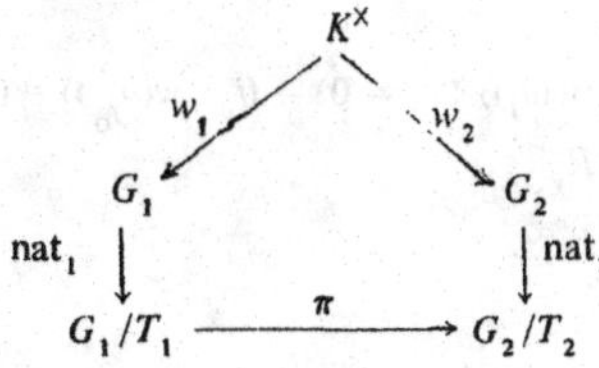

commutates. In this case $u = \text{nat}_1 \cdot w_1 = \text{nat}_2 \cdot w_2 \lneqq w_1, w_2$ is a valuation of K such that $u \leqq w_1 \wedge w_2$. Therefore, $H_i \subseteq T_i$ and it follows $H_i = T_i$, $H_i' = \Delta_i$ and $\nu = (w_1 \wedge w_2)'$. ∎

Now, let p be an irreducible polynomial in $K[X]$, $p \neq X$, and let ν_p be a p-adic valuation of $K(X)$. Let $\{\nu_{p_j} : j \in J'\} = \{\nu_p : p$ an irreducible polynomial in $K[X]$, $p \neq X\}$ and let $\nu_j := \nu_{p_j}$, $j \in J'$,

$$\Omega = \{w_i' : 1 \leqq i \leqq n\} \cup \{\nu_j : j \in J'\},$$
$$R' = \cap R_\nu (\nu \in \Omega).$$

Recall that for a family of valuations Ω' of a field L, Ω' is said to be *weakly independent* provided that for every ν', $\nu \in \Omega'$, $a' \in G^+_{\nu'}$, such that $\delta_{\nu', \nu}(a') = 0$ there exists $x \in B = \cap R_\nu(\nu \in \Omega')$ such that $\nu(x) = 0$, $\nu'(x) \geqq a'$. Moreover, Ω' is called *well centred on* B provided that for every $\nu \in \Omega'$, $a \in G^+_\nu$, there exists $x \in B$ such that $\nu(x) = a$.

6.27. LEMMA (1). *For every valuation w of K and every irreducible polynomial p of $K[X]$, $p \neq X$, $\nu_p \wedge w'$ is a trivial valuation of $K(X)$.*

(2) $R_{w'} \subseteq R_{\nu_X}$ for every valuation w of K.

(3) Ω is weakly independent.

(4) Ω is well centred on R'.

P r o o f. (1) In fact, since ν_p is a rank one valuation, we have either $w' \gneqq \nu_p$ or $w' \wedge \nu_p$ = trivial. But since $\nu_p(X) = 0$, we have $X/p \notin R_{\nu_p}$ and, on the other hand, for $p = a_0 + a_1 X + \ldots + a_m X^m$ (clearly $a_0 \neq 0$) we have

$$w'(X/p) = (1, 0) - w'(p) = (1, 0) - (0, w(a_0)) > 0$$

and it follows $X/p \in R_{w'}$, $w' \ngeqq \nu_p$.

(2) Let $f = f_1/f_2 \in R_{w'}$, $f_1 = a_0 + \ldots + a_n X^n$, $f_2 = b_0 + \ldots + b_m X^m$. We consider the two following cases.

(I) $b_0 \neq 0$. Then $(0, 0) \leqq \min \{(i, w(a_i)) : a_i \neq 0\} - (0, w(b_0))$ and it follows $a_0 \neq 0$ and $\nu_X(f) = 0$. Hence, $f \in R_{\nu_X}$.

(II) $b_0 = 0$. Then $(0, 0) \leqq \min \{(i, w(a_i)) : a_i \neq 0\} - (j_0, w(b_{j_0})) = (i, w(a_i)) - (j_0, w(b_{j_0}))$ and it follows $i \geqq j_0$, $f \in R_{\nu_X}$.

(3) Let $w'_i \in \Omega$ and let us consider at first that for $w'_j \in \Omega$, $(m, \alpha) \in (Z \times G_j)_+$ is such that for $(m, \alpha) = w'_j(g)$, $g \in K[X]$, we have $w'_i \wedge w'_j(g) = (0, 0)$ (i.e. $\delta_{w'_i, w'_j}((m, \alpha)) = 0$). Using 6.26 we obtain

$$(0, 0) = w'_i \wedge w'_j(g) = (w_i \wedge w_j)'(g) = (m, w_i \wedge w_j(a_m)),$$

where $g = \Sigma a_k X^k$, $a_m \neq 0$, $k = m, m+1, \ldots, n$. Hence, $m = 0$, $w_i \wedge w_j(a_0) = 0$. Let $b \in K$ be such that $w_j(b) \geqq \alpha$, $w_i(b) = 0$. Then

$$w'_i(b) = (0, 0), \ w'_j(b) \geqq (m, \alpha).$$

Now, let $v_p \in \Omega$ and let $m \in Z_+$, $p = a_0 + \ldots + a_m X^m$, $a_0 \neq 0$. We set

$$f = ((1/a_0) \cdot p)^m \in X . K[X] + \cap R_{w_i} \ (1 \leqq i \leqq n) = R'.$$

Then $v_p(f) = m \cdot v_p(p) = m$, $w'_i(f) = (0, w_i(1)) = (0,0)$. Finally, let $v_p \in \Omega$ and let us consider an element $(m, \alpha) \in (Z \times G_i)_+$ (in this case $\delta_{w'_i, v_p}((m, \alpha)) = 0$ according (1)). Let $g \in K[X]$ be such that $w'_i(g) = (m, \alpha)$. Since $p \neq X$, we have $v_p(X^{m+1}) = 0$, $w'_i(X^{m+1}) = (m+1, 0) > (m, \alpha)$, $X^{m+1} \in R'$. Let $v_q \in \Omega$, $m \in Z_+$, $q \neq p$, $q = a_0 + \ldots + a_s X^s$, $a_0 \neq 0$. Since $(/w_i(a_0)/)_{1 \leqq i \leqq n} \in A_1$, there exists an element $b \in K$ such that

$$w_i(b) = /w_i(a_0)/, \ i = 1, \ldots, n,$$

and it follows $b \in R'$. In this case $b . q \in R'$ and, moreover,

$$v_q((b \cdot g)^m) = m,$$
$$v_p((b \cdot q)^m) = 0, \ (b \cdot q)^m \in R'.$$

Therefore, Ω is weakly independent.

(4) Let $w'_i \in \Omega$, $(m, \alpha) \in (Z \times G_i)_+$. Then $aX^m \in R'$, where $a \in K$ is such that $w_i(a) = \alpha$ and we have $w'_i(aX^m) = (m, \alpha)$. Let $v_p \in \Omega$, $n \in Z_+$, $p = a_0 + \ldots + a_m X^m$, $a_0 \neq 0$. Let $b \in \cap R_{w_i}$ $(1 \leqq i \leqq n)$ be such that $w_i(b) = /w_i(a_0)/$, $i = 1, \ldots, n$. Then

$$(b . p)^n \in R', \ v_p((b \cdot p)^n) = n,$$

and Ω is well centred on R'. ∎

Now, ve have

$$R' \subsetneqq R_{v_X} \cap \cap R_{v_j} \ (j \in J') = K[X].$$

Let

$$A' = \{(a_1, \ldots, a_n) \in \Pi(\mathbf{Z} \times G_i)_+ : \delta_{w_i, w_j}(a_i) = \delta_{w_j, w_i}(a_j),\ 1 \leqq i,j \leqq n\},$$
$$B' = G(R)_+ .$$

Then for every $(a_i)_i \in A'$ there exists $b \in K(X)$ such that

$$w_i'(b) = a_i,\quad 1 \leqq i \leqq n,$$
$$v_j(b) \geqq 0,\quad j \in J'.$$

Hence, $b \in R'$ and we may set

$$\xi'_*((a_1, \ldots, a_n)) = w_{R'}(b).$$

This definition is correct. In fact, let $b \in R'$ be such that $w_i'(b) = w_i'(b')$, $v_j(b') \geqq 0$, $j \in J'$. Since

$$U(R') = U(\cap R_{w_i'}) = U(\cap R_{w_i}) \subseteq K,$$

we have $b' \cdot b^{-1} \in U(R')$ and it follows $w_{R'}(b') = w_{R'}(b)$.

Hence, we may obtain the approximation theorem

$$\xi' : A' \longrightarrow B'.$$

Then the following proposition holds.

6.28. PROPOSITION. *The approximation theorem* ξ_1 *may be derived from* ξ'.

Proof. For the proof we use 6.23. At first it may be observed that ξ' is represented by a sheaf F'. Let R (R') be the site over which is represented ξ_1 (ξ'). Using the proof of 6.19 it may be observed that R (R') is an $\wedge$-semilattice generated by the set $\{w_i : 1 \leqq i \leqq n\}$ ($\{w_i' : 1 \leqq i \leqq n\}$) and using 6.26 it follows $\mathsf{R} \cong \mathsf{R}'$ as sites. Let F be a sheaf over R which represents $\xi = \xi_1$, i.e.

$$F: \begin{array}{ccc} i & \longmapsto & G_i \\ \downarrow & & \uparrow w_i \\ * & \longmapsto & G(R) \end{array} \qquad\qquad F': \begin{array}{ccc} i & \longmapsto & (\mathbf{Z} \times G_i)_+ \\ \downarrow & & \uparrow w_i \\ * & \longmapsto & G(R')_+ \end{array}$$

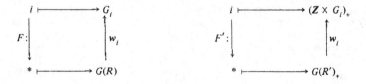

and let $r : F \cdot \mathrm{e} \longrightarrow B$, $r' : F' \cdot \mathrm{e} \longrightarrow B'$ be the natural transformations from the definition of representation, where $\mathrm{e} = (i \longrightarrow *)_{1 \leqq i \leqq n} \in \mathrm{Cov}(*)$. Since $F \cdot \mathrm{e} = B$, $F' \cdot \mathrm{e} = B'$, we have $r = 1_B$, $r' = 1_{B'}$. We define $\alpha : F \longrightarrow F'$. For $w \in \mathrm{R}$, $w \neq *$, we set

$$\alpha_w : G_w \longrightarrow (Z \times G_w)_+$$
$$\alpha_w(a) = (1\, , a)\, .$$

A map α_* is defined as follows. For $a \in G(R)$ there exists the unique $a = (a_1, \ldots, a_n) \in A_1$ such that $\xi_*(a) = a$. Then clearly $((1\, , a_1), \ldots, (1\, , a_n)) \in A'$ and we set

$$\alpha_*(a) = \xi'(((1\, , a_1), \ldots, (1\, , a_n))) \in G(R')_+\, .$$

Then α is a natural transformation, since the diagram

$$\begin{array}{ccc} G_w & \xrightarrow{\alpha_w} & (Z \times G_w)_+ \\ {\scriptstyle w}\uparrow & & \uparrow{\scriptstyle w} \\ G(R) & \xrightarrow{\alpha_*} & G(R')_+ \end{array}$$

commutates. Further, we define a natural transformation $\beta : F' \longrightarrow F$ such that for $w \in \mathrm{R}$, $w \neq *$, we set

$$\beta_w : (Z \times G_w)_+ \longrightarrow G_w$$
$$\beta_w((n, a)) = a\, .$$

A map β_* is constructed as follows. Let $b \in G(R')_+$, then there exists the unique $a' \in A'$ such that $\xi'_*(a) = b$. Let $a = ((m_1\, , a_1), \ldots, (m_n\, , a_n))$. Since $R' \subset K[X]$, we have $m_1 = \ldots = m_n$ and $(a_1, \ldots, a_n) \in A_1$. We set

$$\beta_*(b) = \xi_*((a_1, \ldots, a_n))\, .$$

Again, since $w \cdot \beta_* = \beta_w \cdot w'$, β is a natural transformation and, clearly, $\beta \cdot \alpha = 1$.

Evidently holds $\mathrm{pull}(1, 1) \leqq \mathrm{pull}(\beta, \beta)$ and it follows from 6.23 that ξ_1 may be derived from ξ'. ∎

Let $F : \mathrm{S}^{op} \longrightarrow \mathrm{Set}$ be a presheaf over a site S. Then a many-sorted language L_{S} may be considered such that sorts in L_{S} correspond to objects of S and the only functional symbols are unary functional symbols which correspond to morphisms in S^{op}. In this case S^{op} may be considered as a theory in L_{S} and F is a model of this theory. Hence, we may describe, for example, a relation between two presheaves $F, G : \mathsf{S}^{op} \longrightarrow \mathrm{SET}$ using model-theoretic notions.

For example, it is easy to see that F is a submodel of G if and only if there exists a natural transformation $\alpha : F \longrightarrow G$ which is a monomorphism. Hence, if F_1, F_2 are sheaves over R which represent approximation theorems ξ_1, ξ_2 from 6.11. and 6.12, respectively, it follows that F_2 is a submodel of F_1. In the following theorem we show that there is a more deeper relation ship between F_1 and F_2.

6.29. **THEOREM.** *A model F_1 is an elementary extension of a model F_2.*

Proof. Let $\mu(x_1, \ldots, x_m)$ be a formulae in a language L_{R} such that

$$\mu(x_1, \ldots, x_m) \equiv (\exists x_0)\, \varphi(x_0, x_1, \ldots, x_m)$$

and let $a_1, \ldots, a_m \in F_2$ be values of x_i such that $F_1 \models \mu(a_1, \ldots, a_m)$. (Here we use the following abbreviation: Let x be a variable of a sort ν (i.e. $\nu \in /\mathsf{R}/$), then '$a \in F_2$ is a value of x' means that $a \in F_2(\nu)$ is a value of x.) Without lost of generality φ is assumed to have the following form

$$\varphi(x_0, \ldots, x_m) \equiv f_1(x_0) = x_1 \wedge \ldots \wedge f_m(x_0) \neq x_m .$$

Moreover, we may assume that if in φ subformulas with the same functional symbols appear then these subformulas are in form of negation of equality.

In the following R is assumed to be from the proof of 6.19.

Let V be the sort of x_0 and let for every $K \in /\mathsf{R}/$, f_K be a functional symbol (if it exists) in L_{R} of the sort (V, K) (i.e. $f_K : V \longrightarrow K$ is an arrow in R^{op}). Let

$$J = \{K \in /\mathsf{R}/ : f_K \text{ appears in } \varphi\},$$
$$J_1 = \{K \in J : f_K(x_0) = x_K \text{ appears in } \varphi\},$$
$$J_2 = J - J_1 .$$

Then we may define a map $\rho : J_2 \longrightarrow Z_+$ such that

$$f_K(x_0) \neq x_K^1 \wedge \ldots \wedge f_K(x_0) \neq x_K^{\rho(K)}$$

is the largest subformulae of φ which contains f_K and, moreover,

$$a_K^1 < \ldots < a_K^{\rho(K)}$$

for interpretation of x_K^i. Hence, we may assume that all free variables of φ are of the form x_K, $K \in J_1$, x_R^i, $R \in J_2$, $1 \leqq i \leqq \rho(R)$.

Let for every $K \in /R/$, $v_K = \inf \{w_i : i \in K\}$. The following two cases are considered.

(I) $V \neq *$. In this case we have $v_V \geqq v_K$ for every $K \in J$ and since v_V is a valuation, the set $\{v_K : K \in J\}$ is totally ordered, i.e.

$$v_{K_1} < \ldots < v_{K_m} \leqq v_V ,$$

hence,

$$K_1 \supset \ldots \supset K_M .$$

Let G_i be the value group of v_{K_i}. In this case we have $G_i \cong G_V/H_i$, where H_i is a suitable convex subgroup of G_V, $H_1 \supset \ldots \supset H_m$. Let $b \in G_V = F_1(V)$ be such that

$$f_K^{F_1}(b) \ \ (= F_1(f_K)(b)) = b + H_K = a_K , \ \ K \in J_1 ,$$
$$f_R^{F_1}(b) \neq a_R^i , \ \ R \in J_2 , \ \ 1 \leqq i \leqq \rho(R) .$$

In what follows we frequently identify f_K with its representation $f_K^{F_i}$ in a model F_i, i.e. f_K is the cannonical map $G_V \longrightarrow G_V/H_K$, $G_V^+ \longrightarrow (G_V/H_K)_+$, respectively.

Now, if $J_1 \neq \phi$, we set $c = /b/$, otherwise let c be an element of G_V^+ such that

$$f_{K_1}(c) \neq a_{K_1}^i , \ \ 1 \leqq i \leqq \rho(K_1) .$$

Let $J_2 = \phi$. Then we have

$$f_K(/b/) = /f_K(b)/ = /a_K/ = a_K , \ \ K \in J ,$$

and it follows $F_2 \models \mu(a_K, a_R^i)$. Let $K \in J_2$ be the largest element such that $f_K(c) = a_K^j$ for some j, $1 \leqq j \leqq \rho(K)$. If $K = K_1$ then from the construction of c it follows that $J_1 \neq \phi$, hence $c = /b/$ and since $f_{K_1}(b) \neq a_{K_1}^j$, it follows $f_{K_1}(b) < 0$. But, $v_{K_1} < v_R$ holds for every R and we obtain $f_R(b) < 0$. Therefore, $J_1 = \phi$, a contradiction. Hence, $K \subset K_1$ and $K = K_i$ for some i, $i > 1$. Let $d \in H_{i-1}^+ - H_i$. Then we may find a natural number $n \in \mathbf{Z}_+$ such that

$$f_{K_i}(c + nd) \neq a_{K_i}^j, \quad 1 \leqq j \leqq \rho(K_i),$$
$$f_K(c + nd) = f_K(c), \quad K \supset K_i.$$

Therefore, instead of c we may consider an element $c + nd \in G_V^+$ and after repeating this procedure several times we may find an element $c \in G_V^+$ such that

$$f_K(c) = a_K, \quad K \in J_1,$$
$$f_R(c) \neq a_R^i, \quad R \in J_2, \quad 1 \leqq i \leqq \rho(R).$$

Hence, $F_2 \models \mu(a_K, a_R^i)$.

(II) $V = *$. In what follows, we use a notation from the proofs of 6.19. and 6.11. Since the ring R is a Prüfer ring, for every $K \in J$ there exists a convex directed subgroup H_K of $G(R)$ $(= F_1(*))$ such that $G_K \cong G(R)/H_K$. Let $b \in G(R)$ be such that it satisfies μ. Then $(/w_i(b)/)_{1 \leqq i \leqq n} \in A_2$ and for $a = \xi_{2,*}(/w_i(b)/)_i) \in G(R)_+$ we have

$$w_i(a) = /w_i(b)/, \quad 1 \leqq i \leqq n.$$

Then for $K \in J$ and $i \in K$ we have

$$f_K(a) = \delta_{i,K} \cdot w_i(a) = \delta_{i,K}(/w_i(b)/) = /\delta_{i,K} \cdot w_i(b)/ = /f_K(b)/ = /a_K/ = a_K.$$

Let $K \in J_2$. be such that $f_K(a) = a_K^t$ for some t, $1 \leqq t \leqq \rho(K)$. Let

$$H = \cap H_R^+ (R \in J, \ R \nsubseteq K).$$

If $H \subseteq H_K^+$ then according to 6.7, we have $H_R \subseteq H_K$ for some $R \in J$, $R \nsubseteq K$. In this case $v_R \geqq v_K$ and it follows $R \subseteq K$, a contradiction. Hence, there exists an element $d \in H - H_K$, $d > 0$.

Let $R \subseteq K$ and let i, $1 \leqq i \leqq \rho(R)$, be such that it is the largest index for which there exists $n_R \in Z_+$ with

$$f_R\,(a + n_R d) > a^i_R\,.$$

If such an i does not exist, let $n_R = 1$. Let $n = \max\{n_R : R \subseteq K\}$. Then

$$f_R\,(a + nd) \neq a^i_R\,,\quad R \subseteq K,\quad 1 \leqq i \leqq \rho(R)\,.$$

Moreover, since $nd \in H$, we have

$$f_R\,(a + nd) = f_R\,(a),\quad R \in J,\quad J \nsubseteq K\,.$$

Finally, if $R \subseteq K$, we have $R \in J_2$. In fact, since $f_K\,(a) = a^t_K$ and $/f_K\,(b)/ = f_K(a)$, we obtain $f_K\,(b) = -\,a^t_K < 0$. Then for $R \subseteq K$ we have $f_R\,(b) < 0$ and it follows $R \in J_2$. Now, instead of a we may consider an element $a + nd \in G(R)_+$ and this element satisfies at least one subformula of φ more then the original one. Hence, after repeating this procedure several times we may find an element which satisfies μ. i.e. $F_2 \vDash \mu\,[a_K\,,\,a^i_R\,]$.

Therefore, F_1 is an elementary extension of F_2. ■

It should be observed that even in case the existence of natural transformations $\alpha\,,\,\beta$ between two sheaves $F\,,\,G : \mathbf{C}^{op} \longrightarrow$ SET such that $\alpha : G \longrightarrow F$, $\beta : F \longrightarrow G$, $\beta \cdot \alpha = 1$ it need not follows that F is an elementary extension of G (it follows only that G is a submodel of F).

In fact, let us consider the following example. Let $\mathbf{C}$ be an inf-semilattice generated by a family $\{c_1\,,\,.\,.\,.\,.\,,\,c_n\}$ of elements with the largest element $*$. Then $\mathbf{C}$ is a complete category and we set

$$\mathrm{Cov}_0\,(c) = \{(c' \longrightarrow c)_{c' \in \mathbf{C}'} : \mathbf{C}' \subseteq \mathbf{C}\ ,\ c \in \mathbf{C}'\}\,,$$
$$\mathrm{Cov}_0\,(*) = \{(c_i \longrightarrow *)_{1 \leqq i \leqq n}\}\,.$$

Then $\{\mathrm{Cov}_0\,(c) : c \in \mathbf{C}\ \}$ defines a Grothendieck topology on $\mathbf{C}$ and $\mathbf{C}$ is a site. Let 0 be an element, $0 \notin \mathbf{C}$. We define functors $F\,,\,G$ as follows.

$$G(c) = \{c' \in /\mathbf{C}/ : c' \geqq c\}\,,\quad F(c) = G(c) \cup \{0\}\,,$$
$G(c' \longrightarrow c)$, $F(c' \longrightarrow c)$ are natural insertions.

Then G and F are sheaves over $\mathbf{C}$. In fact, since $\{\mathrm{Cov}_0\,(c) : c \in \mathbf{C}\ \}$ is stable under pullbacks (= inf in $(\mathbf{C}\,,\,\leqq)$), it is neccessery to show only the existence of a solution of a compatible system from a localization from $\mathrm{Cov}_0(c)$ to $F(G)$ and it is easy.

Further, we define

$$\alpha : G \longrightarrow F, \ \beta : F \longrightarrow G$$

such that α_c is an insertion, $\beta_c/G(c)$ = identity, $\beta_c(0) = *$. Then it is easy to see that α and β are natural transformations and $\beta \cdot \alpha = 1$. Further, let f be a morphism $* \longrightarrow c_1$ in $\mathbf{C}^{op}$ and let x_0 (x_1, x_2) be variables of a sort $*$ (c_1). Let

$$\mu(x_1, x_2) = (\exists x_0)\,(f(x_0) \neq x_1 \wedge f(x_0) \neq x_2).$$

Then $F \models \mu(c_1, *)$ but $G \not\models \mu(c_1, *)$. Therefore, F is not an elementary extension of G.

7. REALIZATION OF GROUPS OF DIVISIBILITY

In the theory of *po*-groups the concept of a realization has received a considerable attention (see chapter 1, Fuchs [38], Jaffard [65], Ribenboim [116], Šik [127]) , and enables us to consider a *po*-group as an ordered subgroup of an *o*-product of *o*-groups. In this chapter we deal mostly with the realization of a special *po*-group, groups of divisibility and with applications of these realizations in a theory of rings.

The first result in this direction was obtained by L. Fuchs [39].

7.1. **THEOREM.** *Let an integral domain* A *be an intersection of valuation rings* R_w, $w \in J$. *Then the canonical map*

$$\rho : G(A) \longrightarrow \prod_{w \in J} G(R_w)$$

defined by $\rho(w_A(x)) = (w(x))_{w \in J}$ *is a realization of* $G(A)$.

The proof is straightforward. An immediate consequence of this theorem is a fact that a group of divisibility of an integrally closed integral domain admits a realization. Moreover, this fact may be strengthened in the following way:

Let A be an integral domain with the quotient field K such that

$$\forall x \in K, \quad \forall n \in Z_+ \quad (x^n \in A \Rightarrow x \in A).$$

Then $G(A)$ is semiclosed and according to 1.8, $G(A)$ admits a realization.

The natural question arising here is when the converse of 7.1 holds, i.e. if $\rho : G(A) \longrightarrow \prod_{i \in I} G_i$ is a realization of $G(A)$ then there exists a family of valuation rings $\{R_i : i \in I\}$ such that $G(R_i) = G_i$ and $A = \bigcap_{i \in I} R_i$. J.L.Mott [93] has observed that this does not hold in general.

At first, we deal with the conditions under which the converse of 7.1 as we stated above, holds. The principal ideas are due to J. Ohm [105] and the results are mostly contained in Mott [93]. To begin, we need the following definition.

7.2. DEFINITION. Let G and H be *po*-groups, α a (group) homomorphism of G into H. We say that α is a *v-homomorphism*, providing that

$$\forall\, g_0, g_1, \ldots, g_n \in G,\ (g_0 \in \sup(\inf_G(g_1, \ldots, g_n)))\ \text{implies}$$
$$\alpha(g_0) \in \sup(\inf_H(\alpha(g_1), \ldots, \alpha(g_n))))$$

holds.

The notion of v-homomorphism is due to J.Ohm. and subsumes a couple of other definitions of P.Jaffard [65]. This notion may be translated using the tools of Jaffard's monograph [65] in the following way.

For a directed *po*-group G a map $X \longmapsto X_r$ of the set of all finite subsets of G into $\exp G$ which satisfies

(i) $X \subseteq X_r$,

(ii) $X \subseteq Y_r$ implies $X_r \subseteq Y_r$,

(iii) $\{a\}_r = aG_r = (a)$, $\forall a \in G$,

(iv) $aX_r = (aX)_r$, $\forall a \in G$,

is called an *r-system of* G. The subsets of G in the form X_r are called *r-ideals*. Among r-systems of G there exists a special one called a *v-system* where

$$(x_1, \ldots, x_n)_v = \{x_1, \ldots, x_n\}_v = \bigcap_{\substack{\{x_1,\ldots,x_n\} \subseteq (y) \\ y \in G}} (y) .$$

Then it is easy to see that a homomorphism $\alpha : G \longrightarrow H$ is a v-homomorphism if and only if

$$\alpha((g_1, \ldots, g_n)_v) \subseteq (\alpha(g_1), \ldots, \alpha(g_n))_v$$

for every $g_1, \ldots, g_n \in G$.

Next we collect some immediate properties of the v-homomorphism.

7.3. LEMMA. *(1) Let* $w : K^* \longrightarrow G$ *be a semivaluation of a field* K *and let* $\beta : G \longrightarrow H$ *be a v-homomorphism. Then* βw *is a semivaluation*

(2) A composition of two v-homomorphisms is a v-homomorphism.

(3) Let $\beta : G \longrightarrow H$ *be an o-homomorphism of l-groups* G *and* H. *Then* β

is a v-homomorphism if and only if it is an l-homomorphism.

(4) The projection of an o-product of directed pogroups is a v-homomorphism.

(5) Let $\alpha_i : G \longrightarrow G_i$ *be o-homomorphisms,* $i \in I$, $\alpha = \prod_{i \in I} \alpha_i : G \longrightarrow \prod_{i \in I} G_i$.
Then α *is a v-homomorphism if and only if* α_i *is a v-homomorphism for every* $i \in I$.

Now, let

$$\rho : G(A) \longrightarrow \prod_{i \in I} G_i$$

be a realization of a group of divisibility of a domain A with the quotient field K such that the composition w_i of the following maps

$$K^* \xrightarrow{w_A} G(A) \xrightarrow{\rho} \prod_{i \in I} G_i \xrightarrow{\mathrm{pr}_i} G_i$$

is a semivaluation for every $i \in I$. Then clearly $A = \bigcap_{i \in I} R_{w_i}$, and in this case we say that a family $\{R_{w_i} : i \in I\}$ is *compatible with* ρ. It follows from 7.3, that if ρ is a ν-homomorphism, w_i is a valuation.

7.4. **REMARK.** We note that if A is an integrally closed integral domain (i.e. an intersection of valuation rings) and ρ is a realization of $G(A)$, it is not possible to say that A is an intersection of a family of valuation rings compatible with ρ. In fact, we consider the o-product $G = \mathbf{Z} \times \mathbf{Z}$. On the group G we consider two order relations:

$(x, y)_1 \geqq (0, 0)$ iff $x > 0$ or $x = 0$, $y \geqq 0$.
$(x, y)_2 \geqq (0, 0)$ iff $y > 0$ or $y = 0$, $x \geqq 0$.

Then $(G, {}_1\geqq)$ and $(G, {}_2\geqq)$ are o-groups and the map

$$\rho : G \longrightarrow (G, {}_1\geqq) \times (G, {}_2\geqq)$$

defined by $\rho((x, y)) = ((x, y), (x, y))$ is a realization of G. Since G is an l-group, by 8.1 there exists a Bezout domain A such that $G(A) = G$. If $A = R_1 \cap R_2$ where $\{R_1, R_2\}$ is compatible with ρ, then $R_i = A_{P_i}$ for some prime ideal P_i of A and $G(R_i) \cong {}_o G(A)/H_i$ where $H_i = m(A - P_i)$ (see 2.3). Then $H_1 = \{0\} \times \mathbf{Z}$, $H_2 = \mathbf{Z} \times \{0\}$ and $G(A)/H_i \cong {}_o\mathbf{Z}$, a contradiction.

One may expect that in a case $A = \bigcap_{w \in J} R_w$, the realization ρ created in 7.1, is a ν-homomorphism. That this does not hold has been shown by J.L.Mott [93] who applied a method of using a notion of *w-system* of $G(A)$:

Let $A = \bigcap_{\nu \in J} R_\nu$. Then for $w_A(x_1), \ldots, w_A(x_n) \in G(A)$ we set

$$(w_A(x_1), \ldots, w_A(x_n))_w = \bigcap_{\nu \in J} \{w_A(x) : \nu(x) \geqq \nu(x_i) \text{ for some } i,\ 1 \leqq i \leqq n\}.$$

We assume now that a realization ρ created in 7.1 is a ν-homomorphism. Then the relation $w_A(g_0) \in (w_A(g_1), \ldots, w_A(g_n))_\nu$ implies $\nu(g_0) \in (\nu(g_1), \ldots, \nu(g_n))_\nu$ for every $\nu \in J$ and we obtain

$$(w_A(g_1), \ldots, w_A(g_n))_\nu \subseteq (w_A(g_1), \ldots, w_A(g_n))_w .$$

The converse inclusion always holds. After verifying the reverse implication, we conclude the following proposition.

7.5. PROPOSITION. *Let* $A = \bigcap_{\nu \in J} R_\nu$ *and let* ρ *be a realization created in 7.1. Then* ρ *is a* ν*-homomorphism if and only if the* ν*-system of* $G(A)$ *coincides with the w-system.*

7.6. EXAMPLE. Let $A = k[X, Y]$, where k is a field. Then $A = \bigcap_{i \in I} R_i \cap R_w$ where $R_i = A_{P_i}$ where $\{P_i : i \in I\}$ is the set of minimal prime ideals of A, and w is a valuation of the quotient field of A centred on (X, Y). Then $1 \in (w_A(X), w_A(Y))_\nu - (w_A(X), w_A(Y))_w$ since $w(X) \leqq 1$, $w(Y) \leqq 1$. Thus, ρ is not a ν-homomorphism.

We describe below how one can use the theory of *d*-groups for investigating realizations of groups of divisibility.

7.7. PROPOSITION. *Let* $\rho : G(A) \longrightarrow \prod_{i \in I} G_i$ *be a realization of* $G(A)$. *Then* $A = \bigcap_{i \in I} R_{w_i}$ *where* $\{R_{w_i} : i \in I\}$ *is compatible with* ρ *if and only if* ρ *is a d-homomorphism of* $(G(A), \oplus_A)$ *into* $\prod_{i \in I} (G_i, \oplus_m)$.

<u>Proof</u>. If ρ is a *d*-homomorphism then $\mathrm{pr}_i \cdot \rho$ is a *d*-homomorphism of $(G(A), \oplus_A)$ onto $(G_i, \oplus_m)$ and it follows that $w_i = \mathrm{pr}_i \cdot \rho \cdot w_A$ is a valuation for every $i \in I$ and the set $\{R_{w_i} : i \in I\}$ is a defining family for A.

Conversely, if $\{R_{w_i} : i \in I\}$ is a defining family for A and $w_i = \mathrm{pr}_i \cdot \rho \cdot w_A$, then for $w_A(c) \in w_A(a) \oplus_A w_A(b)$ we have $w_i(c) \geqq \min(w_i(a), w_i(b))$ for every $i \in I$. Thus, we obtain $w_i(c) \in w_i(a) \oplus_m w_i(b)$, and ρ is required to be a d-homomorphism. ∎

7.8. COROLLARY. *A domain A is integrally closed if and only if a d-group $(G(A), \oplus_A)$ admits a d-realization.*

The following proposition may be used to reprove the result mentioned before 7.4.

7.9. PROPOSITION. *Let $\rho : G(A) \longrightarrow \prod_{i \in I} G_i$ be a realization of a group $G(A)$ and let ρ be a ν-homomorphism. Then $\rho : (G(A), \oplus) \longrightarrow \prod_{i \in I} (G_i, \oplus_m)$ is a d-realization for every multivalued addition $\oplus$ on $G(A)$.*

P r o o f. Let $\oplus$ be a multivalued addition on $G(A)$ and suppose that $\gamma \in \alpha \oplus \beta$ for $\alpha, \beta \in G(A)$. Suppose that $\mathrm{pr}_i(\alpha) > \mathrm{pr}_i(\beta)$. Then it is clear that $\gamma \in (\alpha, \beta)_\nu$. Since ρ and pr_i are ν-homomorphisms, it follows that $\mathrm{pr}_i \cdot \rho(\gamma) \in (\mathrm{pr}_i \cdot \rho(\alpha), \mathrm{pr}_i \cdot \rho(\beta))_\nu$ and we have $\mathrm{pr}_i \cdot \rho(\gamma) \geqq \mathrm{pr}_i \cdot \rho(\beta)$. Suppose that $\mathrm{pr}_i \cdot \rho(\gamma) > \mathrm{pr}_i \cdot \rho(\beta)$. Then since $\beta \in \gamma \oplus \alpha$, we have $\mathrm{pr}_i \cdot \rho(\beta) \geqq \mathrm{pr}_i \cdot \rho(\alpha)$, a contradiction. If $\mathrm{pr}_i \cdot \rho(\gamma) < \mathrm{pr}_i \cdot \rho(\alpha)$, we have $\mathrm{pr}_i \cdot \rho(\beta) \geqq \mathrm{pr}_i \cdot \rho(\gamma)$, a contradiction. Thus, $\mathrm{pr}_i \cdot \rho(\gamma) = \mathrm{pr}_i \cdot \rho(\beta)$, and after verifying the other cases we conclude that ρ is a d-homomorphism. ∎

The following couple of propositions shows that many of the well-known concepts of commutative ring theory may be expressed in terms of d-realizations.

7.10. PROPOSITION. *An integral domain A is a Prüfer domain if and only if the canonical map*

$$\rho : (G(A), \oplus_A) \longrightarrow \prod_{H \in M(G(A))} G(A)/H$$

is a d-realization.

P r o o f. If ρ is a d-realization, then $(G(A), \oplus_A)$ is a Prüfer d-group by 5.11, and A is a Prüfer domain by 5.12.1. The converse may be done analogously. ∎

We need this simple notation. For d-groups $G, G_i, i \in I$, we set $G \cong \sum_{i \in I} G_i$

if there is an o-isomorphism ξ of G onto an o-sum of po-groups G_i, $i \in I$, such that ξ is a d-homomorphism and G_i is a totally ordered local d-group for every $i \in I$.

7.11. PROPOSITION. *An integral domain A is a Krull domain if and only if $(G(A), \oplus_A) \cong \Sigma\{G(A)/H : H$ is a maximal element in $\mathrm{M}(G(A))\}$.*

P r o o f. Let A be a Krull domain. Then $A = \cap\{A_M : M$ is a minimal prime ideal of $A\}$ and A_M is a valuation ring. Let H be a maximal element of $\mathrm{M}(G(A))$. By 4.7 and 4.8, there exists a prime ideal M of A such that

$$(G(A_M), \oplus_{A_M}) \cong_d (G(A)/H, \tilde{\oplus})$$

where $\tilde{\oplus}$ is the factor addition on $G(A)/H$. By 2.3, M is minimal and the canonical map $(G(A), \oplus_A) \longrightarrow \prod_H G(A)/H$ is a d-realization (we use the fact $\tilde{\oplus} \subseteq \oplus_m$). Then the rest follows from the approximation theorem for Krull domains. The converse follows similarly. ∎

In the proposition given below we use the notation which is used in Gilmer, Heinzer [74]. Especially, we say that a family of valuation domains $\{R_i : i \in I\}$ is an *irredundant representation of a domain* A, if $A = \bigcap_{i \in I} R_i$, and for every $j \in I$, $A \neq \bigcap_{i \in I, i \neq j} R_i$. Analogously, a d-realization $\rho : G \longrightarrow \prod_{i \in I} G_i$ of a d-group is called *irreducible* if ρ is an irreducible realization of a po-group G. The proof of the following proposition follows directly from 7.7.

7. 12. PROPOSITION. *An integral domain A admits an irredundant representation if and only if the d-group $(G(A), \oplus_A)$ admits an irreducible d-realization.*

7.13. COROLLARY. **(Gilmer, Heinzer [47]).** *Let $G(A)$ be an l-group and let ρ be an irreducible l-realization of $G(A)$. Then A admits an irredundant representation.*

P r o o f. Since ρ is an l-homomorphism, it is a v-homomorphism by 7.3, and ρ an irreducible d-realization of $(G(A), \oplus_A)$ by 7.9. The rest follows from 7.12. ∎

7.14. PROPOSITION. *If $\{R_i : i \in I\}$ is an irredundant representation of a GCD-domain A which is a v-domain, then for every $i \in I$, R_i is an essential (on A)*

valuation domain centred on a maximal ideal of A.

Proof. By 7.12, the canonical map $\rho : G(A) \longrightarrow \Pi G(R_i)$ is an irreducible realization of an l-group $G(A)$. Since A is a ν-domain, the ν-system of $G(A)$ coincides with some w-system of $G(A)$ (see Gilmer [42]) and by 7.5, ρ is a ν-homomorphism. Then by 7.3, ρ is an l-realization of $G(A)$. Now, for every $i \in I$ there exists $\alpha_i \in G(A)_+$ such that $\mathrm{pr}_i \cdot \rho(\alpha_i) > 1$, $\mathrm{pr}_j \cdot \rho(\alpha_i) = 1$ for each $j \in I$, $j \neq i$. From the proof of 1.13 it follows that $\{\alpha_i : i \in I\}$ is a base for $G(A)$, and by 1.12, α_i' is a minimal prime l-ideal of $G(A)$. By 2.3, there exists a maximal ideal M_i of A such that

$$G(A_{M_i}) \cong_o G(A)/\alpha_i'.$$

On the other hand, it is easy to see that $G(R_i) \cong_o G(A)/\alpha_i'$. Thus, A_{M_i} is a valuation domain and M_i = centr of R_i. Therefore, $R_i = A_{M_i}$. ∎

F. Šik [127] and P.Ribenboim [116] introduced several types of l-realization of l-groups, especially, the so-called *reduced, completely regular*, I-, Π'- *realizations*, which were investigated by them in detail Recall that an l-realization $\rho : G \longrightarrow \prod_{i\in I} G_i$ of an l-group G is *reduced*, if for every $i, j \in I$, $i \neq j$, there exists $\alpha \in G$ such that $\rho_i(\alpha) > 1$ (where $\rho_i = \mathrm{pr}_i \cdot \rho$), $\rho_j(\alpha) < 1$; it is called *completely regular*, if for any $i \in I$, $\alpha \in G_+$, such that $\alpha \in \pi(i) = \{\gamma \in G : \rho_i(\gamma) = 1\}$, there exists $\beta \in G_+$ such that $\alpha \wedge \beta = 1$, $\beta \notin \pi(i)$; it is called an *I-realization* if every l-ideal of G is an intersection of a subset of a realizator of ρ; and it is called a Π'-*realization* if a realizator of ρ consists of all minimal prime l-ideals of G. It should be observed that $\{\pi(i) : i \in I\}$ is a realizator of ρ.

Now, if A is a Bezout domain with $A = \bigcap_{i\in I} R_{w_i}$, then the canonical map

$$\rho : G(A) \longrightarrow \prod_{i\in I} G(R_{w_i})$$

is an l-realization. In the following proposition we show some relations between different types of l-realizations of $G(A)$ and a family $\{R_{w_i} : i \in I\}$.

7.15. **PROPOSITION.** *(1)* ρ *is a reduced realization if and only if the set* $\{\mathrm{centr}_A\ w_i : i \in I\}$ *is incomparable with respect to inclusion, where* $\mathrm{centr}_A\ w_i = \{x \in A : w_i(x) > 1\}$.

(2) ρ is a completely regular realization if and only if $\mathrm{centr}_A w_i$ *is a maximal ideal of* A *for every* $i \in I$.

(3) ρ *is a* Π'*-realization if and only if* $\{\mathrm{centr}_A w_i : i \in I\}$ *is the set of maximal ideals of* A.

(4) ρ *is an* I*-realization if and only if for every saturated multiplicative system* S *in* A *there exists* $I_S \subseteq I$ *such that* $A_S = \bigcap_{i \in I_S} R_{w_i}$.

P r o o f. (1) Let ρ be a reduced realization and suppose that $P_i = \mathrm{centr}_A w_i \subsetneq \mathrm{centr}_A w_j = P_j$, $i \neq j$. Since A is a Bezout domain, it follows that $R_{w_i} = A_{P_i} \supsetneq A_{P_j} = R_{w_j}$ and $G(R_{w_i}) \cong_o G(R_{w_j})/H$ for some $H \in O(G(R_{w_j}))$. Then there exists $x \in K$ such that $w_j(x) = \rho_j \cdot w_A(x) < 1$, $w_i(x) > 1$. But, $w_i(x) = w_j(x)H \leqq H$, a contradiction.

Conversely, let $i, j \in I$, $i \neq j$, and let $a \in P_i - P_j$, $b \in P_j - P_i$. Then for $\alpha = w_A(ab^{-1})$ we have $\rho_i(\alpha) = w_i(a) \cdot w_i(b)^{-1} > 1$, $\rho_j(\alpha) = w_j(a) \cdot w_j(b)^{-1} < 1$ and ρ is reduced.

(2) Let ρ be completely regular and suppose that there exists a maximal ideal M of A such that $P_i \subset M$. For $a \in M - P_i$ we have $\alpha = w_A(a) \in \pi(i)$, and by the assumption there exists $\beta = w_A(b) \in G(A)_+$ such that $\alpha \wedge \beta = 1$, $w_i(b) > 1$. Thus $(a, b) = A$, $b \in P_i \subset M$ and it follows $M = A$, a contradiction. Therefore, P_i is a maximal.

Conversely, let P_i be maximal for every $i \in I$ and let $i \in I$, $\alpha \in G(A)_+$, $\alpha \neq 1$, be such that $\alpha \in \pi(i)$. Then for $a \in A$, $w_A(a) = \alpha$, we have $a \notin P_i$ and $(P_i, a) = A$. Hence, for some $b \in P_i$, $b \neq 0$, we have $(a, b) = A$ and $w_A(a) \wedge w_A(b) = 1$, $w_A(b) \notin \pi(i)$.

(3) This follows from 2.3.

(4) Suppose that ρ is an I-realization and let S be a saturated multiplicative system in A. By 2.3, $G(A_S) \cong_o G(A)/H$, where $H = m(S)$. But, A_S is a Bezout domain and it follows that $G(A_S)$ is an l-group. Hence, H is an l-ideal of G and there exists $I_S \subseteq I$ such that

$$H = \bigcap_{i \in I_S} H_i$$

where $\{H_i : i \in I\}$ is a realizator of ρ. Then

$$A_S = \bigcap_{i \in I_S} R_{w_i}.$$

For, if $ab^{-1} \in A_S$, $a \in A$, $b \in S$, then $w_A(b) \in H$ and $w_i(b) = 1$ for every $i \in I_S$. Hence, $ab^{-1} \in R_{w_i}$, $i \in I_S$. Conversely, if $x \in R_{w_i}$ for every $i \in I_S$, it follows $w_A(x)H \geqq H$. Hence, there exists $w_A(c) \in H_+$ such that $w_A(c) \geqq w_A(b)$. Then $w_A(xb^{-1}) \geqq 1$, $w_A(cb^{-1}) \geqq 1$, $w_A(cb^{-1}) \in H$ and we have $cb^{-1} \in S$. Thus, $x = (cxb^{-1}) \cdot (cb^{-1})^{-1} \in A_S$. The converse implication may be proved similarly. ∎

Now, from the fact that not every completely regular *l*-realization is a Π'-realization we obtain the following proposition.

7.16. PROPOSITION. *There exists a Bezout domain A such that the set $\{A_M : M$ is a maximal ideal of $A\}$ is not an irredundant representation of A.*

Proof. Let G be an *l*-group such that there exists a completely regular *l*-realization $\rho : G \longrightarrow \prod_{i\in I} G_i$ which is not a Π'-realization. Hence, there exists a minimal prime *l*-ideal H of G such that $H \notin \{\pi(i) : i \in I\}$. Let A be a Bezout domain such that $G(A) = G$ (see 8.1). Then $M = A - m^{-1}(H)$ is a maximal ideal of A and $G(A_M) \cong_o G/H$, $A = \bigcap_{i\in I} A_{M_i}$ for a set of maximal ideals $M \notin \{M_i : i \in I\}$. ∎

8. JAFFARD – OHM's THEOREM

8.1. THEOREM. *If* G *is an l-group then there exists a Bezout domain* A *such that* $G(A) \cong_o G$.

The evolution of this result, being one of the most important from the theory of groups of divisibility, comprises a period of at least 30 years. The first significant step in this sphere was done by W. Krull [79] who observed that for any o-group G there exists a valuation domain R such that $G(R) \cong_o G$. The proof of this theorem is based on the notion of the *group algebra of* G *over a field.* Namely, let k be an arbitrary field and let $k[G]$ be the set of functions from G into k which are finitely nonzero. Then $k[G]$ is an integral domain with identity if, for $f, g \in k[G]$, $f + g$, $f \cdot g$ are defined

$$(f+g)(\alpha) = f(\alpha) + g(\alpha),$$
$$(f \cdot g)(\alpha) = \sum_{\beta+\gamma=\alpha} f(\beta) \cdot g(\gamma),$$

where the notation $\sum_{\beta+\gamma=\alpha}$ signifies that the summation is taken over all pairs (β, γ) of elements of G such that $\alpha = \beta + \gamma$. If $f \in k[G]$ and $\{\alpha_1, \ldots, \alpha_n\}$ is the set of elements of G on which f does not vanish, then f can be identify with the 'polynomial' $\Sigma f(\alpha_i) X^{\alpha_i}$. Then operations in $k[G]$ correspond to the usual operations among the associated polynomials. We denote by $k(G)$ the quotient field of $k[G]$ and let

$$w : k[G] \longrightarrow G$$

be defined by

$$w(a_1 X^{\alpha_1} + \ldots + a_m X^{\alpha_m}) = \min\{\alpha_1, \ldots, \alpha_m\}.$$

Then it is easy to see that w has the properties

(i) $\quad w(f \cdot g) = w(f) + w(g)$,

(ii) $\quad w(f - g) \geqq \min\{w(f), w(g)\}$,

and it follows that w may be extended onto a valuation (denoted by the same symbol) w of a field $k(G)$.

Jaffard, following the major step of Krull's proof, first constructed a domain A with an l-group G as its group of divisibility. Then Ohm [105] astutely observed that the domain A in this construction is necessarily a Bezout domain. J.L.Mott [93] has observed that Kaplansky had made the same observation 12 years earlier in the unpublished part of his dissertation at Harvard University.

We show the two version of the proof of 8.1.

Proof (Jaffard). Let $k[G]$ be the group algebra of an l-group G over a field k. For $f = a_1X^{\alpha_1} + \ldots + a_nX^{\alpha_n} \in k[G]$ we set

$$\operatorname{supp}(f) = \{\alpha_1, \ldots, \alpha_n\},$$
$$w(f) = \inf_G \{\alpha : \alpha \in \operatorname{supp}(f)\}.$$

Then $w(f \cdot g) = w(f) + w(g)$ for $f, g \in G$. In fact, without the lost of generality we may assume that $w(f) = w(g) = 0$, since otherwise we may divide the polynomials g and f by $X^{w(g)}$, $X^{w(f)}$, respectively. Let

$$f = a_1X^{\alpha_1} + \ldots + a_mX^{\alpha_m}, \quad g = b_1X^{\beta_1} + \ldots + b_nX^{\beta_n},$$
$$f \cdot g = c_1X^{\gamma_1} + \ldots + c_pX^{\gamma_p},$$
$$\alpha_1 \wedge \ldots \wedge \alpha_m = \beta_1 \wedge \ldots \wedge \beta_n = 0.$$

We need an auxiliary fact.

8.1.1. LEMMA. *Let $(G,+)$ be an l-group and let $\alpha_1, \ldots, \alpha_n \in G_+$. Then $\alpha_1 \wedge \ldots \wedge \alpha_n = 0$ if and only if for every $\xi \in G$ there exist i, $1 \leqq i \leqq n$, and an element $\xi' \in G$ such that*

$$0 < \xi' \leqq \xi, \quad \alpha_i \wedge \xi' = 0.$$

Proof. Let $\alpha_1 \wedge \ldots \wedge \alpha_n = 0$. We set $\xi_0 = \xi$, $\xi_i = \xi \wedge \alpha_1 \wedge \ldots \wedge \alpha_i$. Then $\xi = \xi_0 \geqq \xi_1 \geqq \ldots \geqq \xi_n = 0$. Let i be the first index such that $\xi_i = 0$. Then we set $\xi' = \xi_{i-1}$. Conversely, we admit that

$$\xi = \alpha_1 \wedge \ldots \wedge \alpha_n > 0$$

and the conditions of 8.1.1 are satisfied. Then for any $\xi' \in G$ such that $0 < \xi' \leqq \xi$ we have $\alpha_i \wedge \xi' = \xi' > 0$, $i = 1, \ldots, n$, a contradiction.

Now, let $\xi \in G$, $\xi > 0$, be an arbitrary element. Applying the 8.1.1 twice, there exist $\delta \in G$ and indices i, j such that $\xi \geqq \delta > 0$, $\alpha_i \wedge \delta = \beta_j \wedge \delta = 0$. We may suppose that indices m', n' are selected such that

$$\alpha_t \wedge \delta = 0 \quad \text{iff} \quad t \leqq m' \quad (1 \leqq m' \leqq m),$$
$$\beta_s \wedge \delta = 0 \quad \text{iff} \quad s \leqq n' \quad (1 \leqq n' \leqq n).$$

We set $f = f_1 + f_1'$, $g = g_1 + g_1'$, where

$$f_1 = a_1 X^{\alpha_1} + \ldots + a_{m'} X^{\alpha_{m'}}, \quad g_1 = b_1 X^{\beta_1} + \ldots + b_{n'} X^{\beta_{n'}}.$$

Then $f \cdot g = f_1 g_1 + t$, where $t = f_1 g_1' + f_1' g_1$. Since for every $\alpha \in \operatorname{supp}(f_1 g_1)$, $\beta \in \operatorname{supp}(t)$, $\alpha \wedge \beta = 0$, $\beta \wedge \delta \neq 0$, hold, we have $\operatorname{supp}(f_1 g_1) \subseteq \operatorname{supp}(f \cdot g)$. Let $\eta \in \operatorname{supp}(f \cdot g)$. If $\xi > 0$, there exists δ' such that $0 < \delta' \leqq \xi$, $\delta' \wedge \eta = 0$. Using 8.1.1 we obtain $\gamma_1 \wedge \ldots \wedge \gamma_p = 0$. Since

$$\operatorname{supp}(f + g) \subseteq \operatorname{supp}(f) \cup \operatorname{supp}(g),$$

we have $w(f + g) \geqq w(f) \wedge w(g)$ and w may be extended on a semivaluation of a field $k(G)$ with a group G by $w(f/g) = w(f) - w(g)$. Clearly, A_w is a Bezout domain.

Proof (Ohm). Let $\rho : G \longrightarrow \prod_{i \in I} G_i$ be an l-realization of G and let $\rho_i = \operatorname{pr}_i \circ \rho$, $i \in I$. Let k be an arbitrary field and let $K = k(\{X_g : g \in G\})$ where X_g is an indeterminate over k. Define maps $w_i : K \longrightarrow G_i$

$$w_i(c X_{g_1}^{n_1} \ldots X_{g_r}^{n_r}) = n_1 \rho_i(g_1) + \ldots + n_r \rho_i(g_r),$$
$$w_i(x) = \inf_{G_i} \{w_i(m) : m \text{ are the distinct monomials appearing in } x\},\ x \in k[\{X_g\}],$$
$$w_i(x/y) = w_i(x) - w_i(y),\ x/y \in K.$$

Then for $x \in K$ we set

$$w(x) = \rho^{-1}((w_i(x))_{i \in I}) \in G.$$

It is easy to see that w_i are valuations of K and that w is a semivaluation of K.

Let $A = A_w$. Then A is an integral domain with the quotient field K, and $G(A) = = w(K^*)$. Let $g \in G$, then $w(X_g) = \rho(g) \in w(K^*)$ and $\rho(G) \subseteq w(K^*)$. On the other hand, $w(K^*)$ is a sublattice of $\prod_{i \in I} G_i$ generated by $\rho(G)$, hence, $\rho(G) = w(K^*)$ and $G(A) \cong_o G$. If we select for $x, y \in K$, $x = x_1/x_2$, $y = y_1/y_2$, $x_i, y_i \in k[\{X_g\}]$, an exponent $m >$ the degree of X_0 in x_1 and if we set $z = x + X_0^m y$ then

$$w(z) = w(x_1) \wedge w(y_1) - w(y_2) = w(x) \wedge w(y),$$

and $(x, y) = (z)$ in A. Therefore, A is a Bezout domain.

8.2. **REMARK.** For the domain A constructed in the second proof the following holds: $\oplus_A = \oplus_m$ in G $(= G(A))$.

In fact, let $w(c) \in w(a) \oplus_m w(b)$ for some $a, b, c \in K$. Then $w(a) \wedge w(b) = = w(c) \wedge w(a) = w(c) \wedge w(b)$. Hence, there exist $m_1, m_2 \in Z_+$ such that

$$w(a) \wedge w(b) = w(a + X_0^{m_1} b), \quad w(a) \wedge w(c) = w(c + X_0^{m_2} a).$$

Hence, there is an $u \in U(A)$ such that

$$c = a(u - X_0^{m_2}) + u X_0^{m_1} b .$$

Since $w(X_0^{m_i}) = 0$, it follows that $uX_0^{m_i} \in U(A)$. Now, let $u = p/q$ where $p = = p_1 + \ldots + p_t$, $q = q_1 + \ldots + q_s$, $p, q \in k[\{X_g\}]$ and p_i (q_j) are the distinct monomials appearing in p (q). Since $w(u) = 0$, it follows that

$$w_i(p_1) \wedge \ldots \wedge w_i(p_t) = w_i(q_1) \wedge \ldots \wedge w_i(q_s)$$

for every $i \in I$. Then

$$w_i(u - X_0^{m_2}) = w_i(p_1) \wedge \ldots \wedge w_i(p_t) \wedge w_i(-q_1) \wedge \ldots \wedge w_i(-q_s) - \\ - w_i(q_1) \wedge \ldots \wedge w_i(q_s) = 0$$

and $u - X_0^{m_2} \in U(A)$. Thus

$$w(c) \in w(a) \oplus_A w(b) .$$

Since the converse inclusion always holds, we have $\oplus_A = \oplus_m$.

Ohm's construction of an integral domain with a prescribed l-group as a group of divisibility enables us to construct the group of divisibility of some well-known ring constructions. We illustrate it with the case of Kronecker's function ring A^b of a domain A with the quotient field K where

$$A^b = \{f/g : f, g \in A[X], \ c(f)_b \subseteq c(g)_b\} \cup \{0\}$$

where $c(f)$ is an ideal of A generated by the coefficients of f and $J_b = \bigcap_{w \in W} J \cdot R_w$, where W is the set of valuations of K which are positive on A, for every ideal J of A.

At first, according to Gilmer [42], we have

$$A^b = \bigcap_{w \in W} R_{w'}$$

where w' is the canonical extension of $w \in W$ on the valuation of a field $K(X)$. We consider the following diagram, where ρ is the realization from 7.1 and $\nu = \prod_{w \in W} w'$.

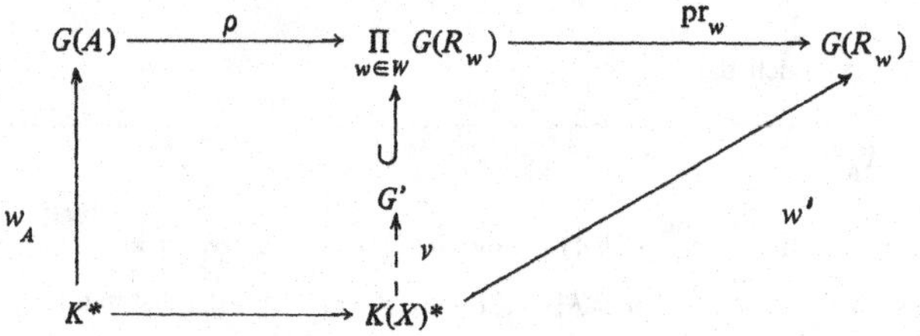

Then G' is a group of divisibility of A^b and, clearly,

$$G(A) \subseteq G' \subseteq G(A)_u$$

where $G(A)_u$ is an l-ideal of $\prod_{w \in W} G(R_w)$ generated by $G(A)$. Since A^b is a Bezout domain (Gilmer [42]), G' is an l-group and $G' = G(A)_u$. Thus, we obtain the following proposition.

8.3. **PROPOSITION.** *If $\{R_w : w \in W\}$ is the family of valuation domains of K containing A, then for the Kronecker function ring A^b of A, $G(A^b)$ is the l-ideal of $\prod_{w \in W} G(R_w)$ generated by $G(A)$.*

We show another such construction utilizing an analogy of Krull's method for constructing a valuation with a pre-assigned group. This construction is due to M. Griffin [51] and enables us to construct a Krull-type ring for an l-group sytisfying some conditions.

Firstly, we repeat some of the facts, which were introduced by Griffin. A domain A with the quotient field K is called a *ring of Krull type* provided that there exists a defining family $\{R_w : w \in J\}$ for A such that

(1) $R_w = A_{P(w)}$ where $P(w) = \mathrm{centr}_A\, w$, $w \in J$,

(2) $\forall\ x \in K^*$, $w(x) = 0$ for all but finite many $w \in J$.

Then M.Griffin [51] showed that for any ring of Krull type A, the group $G(A)$ may be embedded into ar l-group G such that each positive element of G is greater than only a finite number of pairwise disjoint elements of G (in this case, G is said to satisfy *Conrad's (F)-conditions*), in such a way that each element of G is the infimum of a finite number of elements of $G(A)$.

The following example, due to Griffin, shows that given an l-group G satisfying the Conrad's (F)-conditions it is possible to construct a Krull-type ring with G as a group of divisibility.

8.4. EXAMPLE. Let G be an l-group satisfying the Conrad's (F)-conditions. Then there exists an irreducible l-realization

$$\rho : G \longrightarrow \prod_{i \in J} G_i .$$

Let ρ_i be the composition of ρ with the ith projection map, then we say a subset $S \subseteq G$ is called *semi-well-ordered* if $\rho_i(S)$ is well ordered for every $i \in J$, and it is zero for almost all $i \in J$.

Let k be a field and let

$$k[[G]] = \{a = \sum_{g \in G_+} a_g X^g : \mathrm{supp}(a) \text{ is semi-well-ordered}\} .$$

For $a = \sum_g a_g X^g$, $b = \sum_g b_g X^g$ we set $c = \sum_g c_g X^g = a \,.\, b$, $d = \sum_g d_g X^g = a + b$ where

$$c_g = \sum_{r+s=g} a_r \cdot b_s,\quad d_g = a_g + b_g .$$

It is possible to show that the summation is finite, and $\mathrm{supp}(a \cdot b)$ is semi-well-ordered.

Now, we define a map w from $k[[G]]$ into G_+. We distinguish two cases in defining $w(a)$, $a \in k[[G]]$.

I) Either $a = 0$ or a contains a term $a_0 X^0$; in the first case we set $w(a) = w(0) = \infty$ and $w(a) = 0$ in the second one.

II) There is no constant term in the formal power series of a. Let $i_1, \ldots, i_n$ be the indices of J such that $\rho_i(\text{supp}(a)) \neq 0$, and let $\rho_{i_t}(g_t)$ be the first element of $\rho_{i_t}(\text{supp}(a))$. Then we set

$$w(a) = g_1 \wedge \ldots \wedge g_n .$$

Then it is possible to show that $w(a \cdot b) = w(a) + w(b)$ and $w(a + b) \geqq w(a) \wedge w(b)$. Thus, w may be extended onto a semi-valuation of the quotient field K of $k[[G]]$ with the value group G. Let $A = A_w$ and let

$$w_i = \rho_i \cdot w : K \longrightarrow G_i .$$

Then w_i is a valuation of K and $\{R_{w_i} : i \in J\}$ is a defining family for A. Using Krull's approximation theorem 6.9 (1) for l-group G it is possible to show that A is a Krull-type ring.

We note that at the present time, *group rings* (= group algebras over a groups and ring) and *semigroup rings*, including their divisibility properties, are being intensively studied (see Gilmer, Parker [49], Parker [107], Connell [30]). We show in this chapter several results which describe the conditions in order that the group of divisibility of a semigroup ring $A[S]$ is an l-group. For a more detail investigation of divisibility properties including the results presented here, see Gimer, Parker [49].

The following theorem is an analogy of a theorem due to Nagata (see 9.16).

8.5. THEOREM. *Let* N *be a multiplicative system of a domain* A *and let* $T = \{x \in A : w(x \cdot y) = w(x) \vee w(y)$ *for all* $y \in N\}$, *where* $w = w_A$. *Assume that the following conditions are satisfied.*

(1) For every $x, y \in N$ *there exists* $w(x) \vee w(y)$ *in* $G(A)$.

(2) Each nonzero element of A *can be expressed as the product of an element of* N *and an element of* T.

If A_N *is a GCD-domain, then* A *is a GCD-domain.*

Proof. Let $w = w_{A_N}$. The following diagram commutates (see 2.3).

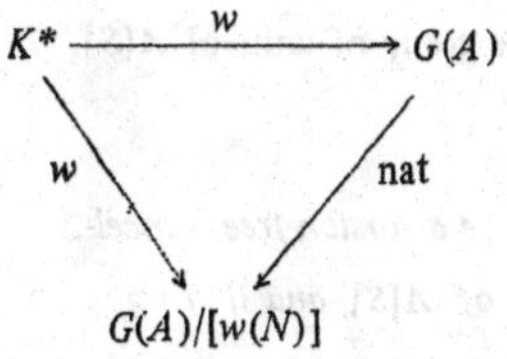

To show that A is a GCD-domain we need to prove that $G(A)$ is an l-group. Let $b, c \in A$ and let $b = n_1 t_1$, $c = c_2 t_2$ for some $n_i \in N$, $t_i \in T$. According to condition (1) we may consider $\alpha = w(n_1) \vee w(n_2) \in G(A)$. Since $G(A_N)$ is an l-group, we have

$$w(b) \vee w(c) = w(t_1) \vee w(t_2) = w(t)$$

for some $t \in T$. We show that $w(b) \vee w(c) = \alpha + w(t)$. By doing so, we prove that

$$(*) \quad a, b \in A, \; n \in N, \; t \in T, \; w(a \cdot n) = w(t \cdot b) \text{ imply } w(a) \geqslant w(t).$$

In fact, since $w(n) \leqslant w(a \cdot n) = w(t \cdot b)$ and $w(t) \leqslant w(t \cdot b)$ we can see that $w(t \cdot n) = w(t) \vee w(n) \leqslant w(t \cdot b)$ and it follows $w(n) \leqslant w(b)$. Then $w(a \cdot b) \geqslant$ $\geqslant w(a \cdot n) = w(t \cdot b)$ and it follows $w(a) \geqslant w(t)$.

Now, since $w(t_1) + H \leqslant w(t) = w(t) + H$, where $H = [w(N)]$, there exist $s_1, s_2 \in N$, $a \in A$ such that $w(t_1 s_1 a) = w(t s_1)$ and according to (*), we obtain $w(t_1) \leqslant w(t)$. Hence,

$$w(b) = w(n_1) + w(t_1) \leqslant \alpha + w(t),$$
$$w(c) = w(n_2) + w(t_2) \leqslant \alpha + w(t).$$

Further, let $w(x) \in G(A)$ be such that $w(x) \geqslant w(b), w(c)$, and let $x = m \cdot s$; $m \in N$, $s \in T$. Then $w(x) \geqslant w(n_i)$, $w(n_i) \vee w(s) = w(n_i \, . \, s)$ and according to (*) it follows $w(n_i) \leqslant w(m)$, $\alpha \leqslant w(m)$. Further, $w(s) = w(m \, . \, s) \geqslant w(b) \vee w(c) = w(t)$ and, again according to (*), we obtain $w(s) \geqslant w(t)$.

$$\alpha + w(t) \leqslant w(m) + w(s) = w(x)$$

and it follows $\alpha + w(t) = w(b) \vee w(c)$. Therefore, $G(A)$ is an l-group and A is a GCD-domain. ∎

The following lemma allows us to describe completely the group of units of $A[S]$ for special S.

8.6. **LEMMA.** *Let A be an integral domain and let S be a torsion-free, cancellative, additive semigroup. If f and g are nonzero elements of $A[S]$ and if $f \cdot g$ is a monomial, then f and g are monomials. Moreover,*

$$U(A[S]) = \{u \cdot X^s : u \in U(A) \text{ and } s \text{ has an additive inverse in } S\}.$$

Proof. Since S is torsion-free and cancellative, it admits a total ordering $\leqslant$ compatible with the semigroup structure. Let

$$f = f_1 X^{s_1} + \ldots + f_n X^{s_n}, \quad g = g_1 X^{t_1} + \ldots + g_n X^{t_m}$$

where $s_1 < \ldots < s_n$, $t_1 < \ldots < t_m$. Because A is an integral domain and S is cancellative, it is clear that $f \cdot g$ is not a monomial if either $n > 1$ or $m > 1$. ∎

Without any proof we state the following theorem (for some analogy, see 9.18).

8.7. **THEOREM.** *Let A be an integral domain with the quotient field K and let S be a torsion-free, cancellative, additive semigroup with zero. Then the following conditions are equivalent.*

(1) $A[S]$ is a GCD-domain.

(2) A and $K[S]$ are GCD-domains.

For the proof which requires several further results see Gilmer, Parker [49].

The following proposition solves the first part of our question for group rings.

8.8. **PROPOSITION.** *If A is a GCD-domain and G is a torsion-free group then $A[G]$ is a GCD-domain.*

Proof. We show that for any two principal ideals $f \cdot A[S]$, $g \cdot A[S]$ their intersection is a principal ideal too, i.e. $G(A[S])$ is an l-group. Let $f = \sum_{i=1} f_i X^{s_i}$, $g = \sum_{i=1} g_i X^{t_i}$ and let H be the subgroup of G generated by the set $\{s_1, \ldots, s_n, t_1, \ldots, t_m\}$. Since elements of H are of the form

$$k_1 s_1 + \ldots + k_n s_n + r_1 t_1 + \ldots + r_m t_m = b_1 h_1 + \ldots + b_u h_u ;$$

where $k_i r_j \in Z$ and $h_i \in \{s_1, \ldots, t_m\}$, $u \leqslant n+m$, $b_j \in Z$, we obtain that $A[H]$ is a quotient ring of a polynomial ring in u indeterminate over A. Hence, $A[H]$ is a GCD-domain. Further, it may be observed without any difficulty that $A[G]$ is a free $A[H]$-modul. So let $\{y_\lambda\}$ be a free modul basis for $A[G]$ over $A[H]$. If $fA[H] \cap gA[H] = hA[H]$, then

$$fA[G] \cap gA[G] = f(\sum_\lambda y_\lambda A[H]) \cap g(\sum_\lambda y_\lambda A[H]) =$$
$$= (\sum_\lambda fA[H]y_\lambda) \cap (\sum_\lambda gA[H]y_\lambda) = \sum_\lambda (fA[H] \cap gA[H])y_\lambda =$$
$$= \sum_\lambda hA[H]y_\lambda = h \cdot A[G],$$

and $A[G]$ is a GCD-domain. ∎

The following theorem shows completely the problem of determining the necessary and sufficient conditions so that a semigroup ring should be a GCD-domain.

8.9. THEOREM. *Let A be an integral domain and let S be an additive semigroup with zero. Then the following conditions are equivalent.*

(1) $A[S]$ is a GCD-domain.

(2) A is a GCD-domain and S is a torsion-free, cancellative GCD-semigroup, i.e. for any $a, b \in S$, $(a+S) \cap (b+S) = c+S$ for some $c \in S$.

P r o o f. (1) ⇒ (2). Let $A[S]$ be a GCD-domain. Then clearly S is torsion--free and cancellative and, according 8.7, A is a GCD-domain. If $a, b \in S$, then $(X^a) \cap (X^b)$ is a principal ideal of $A[S]$ and since $X^{a+b} \in (X^a) \cap (X^b)$, it follows from by 8.6 that

$$(X^a) \cap (X^b) = (X^c)$$

for some $c \in S$. But $(X^t) = X^t . A[S] = A[t+S]$ and it follows

$$(a+S) \cap (b+S) = (c+S) .$$

The proof of the implication (2) ⇒ (1) requires several further results and we omit it here. ∎

At the end of this section we investigate the principal role which l-groups play in the theory of groups of divisibility. It should be observed that one of the most basic topics in divisibility theory is the problem concerning the notion of the greatest common divisor (i.e. the infimum in a group of divisibility). It concerns with the methods for adjoining infimums in case they are missing in a group of divisibility and, moreover, with determining the exact conditions under which such an extension is possible. The problem of the existence of such an extension was originally solved by Krull who defined the *Kronecker function ring* in order to study the arithmetic of integral domains. The principal result concerning this construction is the fact that the Kronecker function ring of an integrally closed domain is Bezout (see 8.3) and, in this case, any group of divisibility of an integrally-closed integral domain A may be inserted in an l-group, namely

$$G(A) \longrightarrow G(A^b) .$$

It should be observed that the existence of such an insertion follows directly from the existence of a realization of a semi-closed *po*-group (see 1.8 and 7.1).

The remainder of this chapter is devoted to investigating the existence of such a minimal insertion. Roughly speaking, the best situation occurs if the category of l-groups with l-homomorpisms is a full reflective subcategory of a sufficiently large category containing at least groups of divisibility of integrally closed integral domains.

The method we use here is a slight modification of Aubert's method in [2]. Aubert essentially used the notion of a Lorenzen group which is a purely multiplicative transcription of a Kronecker function ring originally introduced by Lorenzen [74].

The best approximation of a group of divisibility for our purposes is the notion of a *d-group* (any group of divisibility is a d-group) and this is the reason why the category we are interested in will contain special d-groups only.

In chapter 7 we introduced the notion of a r-system in a directed *po*-group G. To investige the theory of Lorenzen groups we need some further facts from the theory of r-systems. The best descriptions of this theory are Aubert's preprint [2] and Jaffard's monograph [65].

We now investigate some of the most essential technical details.

Let (G_1, r_1) and (G_2, r_2) be two r-systems on d-groups G_1, G_2, respectively. (Clearly, r-system concerns, in general, the *po*-structure of G only). A *(r_1, r_2)-morphism from (G_1, r_1) into (G_2, r_2)* is a map

$$\varphi : G_1 \longrightarrow G_2$$

such that

(1) φ is a d-homomorphism ;

(2) $\varphi(X_{r_1}) \subseteq (\varphi(X))_{r_2}$ for any finite set X in G_1.

For any directed *po*-group G there is a unique coarse r-system, the *t-system*, such that for any finite subset X of G_+ we have

$$X_t = \bigcup_{\substack{Y \subseteq X \\ Y \text{ finite}}} Y_v$$

(for a definition of the v-system, see chapter 7). In a case where G is an l-group, it is easy to see that the set of t-ideals of G_+ coincides with the set of filters in G_+. Moreover, the following simple lemma holds.

8.10. LEMMA. *Let G and H be l-groups and let $\varphi : G \longrightarrow H$ be a map. Then φ is an l-homomorphism if and only if φ is a (t, t)-morphism from $(G, \oplus_m)$ into $(H, \oplus_m)$.*

Proof. Let φ be an l-homomorphism. Then according 3.14, φ is a d-homomorphism from $(G, \oplus_m)$ into $(H, \oplus_m)$. Moreover, if $X = \{x_1, \dots, x_n\} \subset G_+$, then

$$X_t = \{x \in G_+ : x \geqslant x_{i_1} \wedge \dots \wedge x_{i_k} \text{ for some } 1 \leqslant i_1, \dots, i_k \leqslant n\}$$

and it follows $\varphi(X_t) \subseteq (\varphi(X))_t$. Hence, φ is a (t, t)-morphism. The converse follows directly from 3.14. ∎

For any directed d-group G we may construct a special r-system, called the d-system, such that for any finite subset X of G_+ we let X_d be an m-ideal of G_+ generated by X, i.e.

$$X_d = \{x \in G_+ : x \in a_1 x_1 \oplus \dots \oplus a_m x_m \text{ for some } x_i \in X,\ m \geqslant 1\} \supseteq X.$$

The following lemma is an analogy of 8.10.

8.11. LEMMA. *Let $(G, \oplus)$ and $(H, \oplus')$ be directed d-groups and let $\varphi : G \longrightarrow H$ be a map. Then φ is a d-homomorphism if and only if φ is a (d, d)-morphism.*

Moreover, in case $G = (G, \oplus_m)$ it is easy to see that the t-system is identical to the d-system. Clearly, if $X = \{x_1, \ldots, x_n\}$ and $x \in X_t$ then

$$x \geqslant z = x_{i_1} \wedge \ldots \wedge x_{i_k} \in x_{i_1} \oplus_m \ldots \oplus_m x_{i_k} \subseteq X_d$$

and it follows $x = z \cdot g \in X_d$ for some $g \in G_+$. The converse inclusion may be done analogously.

Further, let A be an integral domain and let $G = (G(A), \oplus_A)$. It may be proved without difficulty that there is a bijection between the d-ideals of G_+ and the ideals of a domain A. An ideal J of A corresponds in this bijection to an m-ideal (= d-ideal) $w_A(J)$ of G_+.

For a directed po-group G with an r-system r we say that G is *r-closed* if

$$X_r : X_r \subseteq G_+$$

for any finite set X of G_+ and where

$$X_r : X_r = \{x \in G : x \,.\, X_r \subseteq X_r\}.$$

If $G = (G(A), \oplus_A)$ then G is d-closed if and only if A is integrally closed in its quotient field. For a general directed d-group $G = (G, \oplus)$ we can only prove that if G is d-closed then G_+ is integrally closed in G (see chapter 5). In fact, let

$$x^{n+1} \in a_n x^n \oplus \ldots \oplus a_0 ; \; a_i \in G_+ .$$

Then $x^{n+1} \in \{1, x, \ldots, x^n\}_d = X_d$ and it follows $x \cdot X_d \subseteq X_d$. Hence, $x \in$ $\in X_d : X_d \subseteq G_+$ and $x \in G_+$.

Now, to the given r-system of a directed d-group G we can associate another ideal system in G which is denoted by r_a and which is determined such that

$$X_{r_a} = \{x \in G : xY_r \subseteq X_r \times Y_r \text{ for a finite } Y \subset G\}.$$

(For the definition of a multiplication see chapter 12.) The principal property of r_a-system is that the monoid of finitely-generated r_a-ideals satisfies the cancellation law and hence possesses a group of a quotients $L_r(G)$ (for the proof see Jaffard [65]). This group may be ordered such that

$$L_r(R)^+ = \left\{ \frac{X_{r_a}}{Y_{r_a}} \in L_r(R) : X_{r_a} \subseteq Y_{r_a} \right\}$$

and is as such called the *Lorenzen r-group of* G. Moreover, with respect to the ordering, $L_r(G)$ is an l-group which contains G as an ordered subgroup. An insertion

$$\alpha : G \longrightarrow L_r(G)$$

is such that for $g \in G$ we let $\alpha(g) = \{g\}_{r_a} = g \cdot G_+ \in L_r(G)$. Hence, this construction provides the infimums which are missing in G and we show that it does it in the most economical way.

For this purpose we introduce several notations. Let $\boldsymbol{D}$ be the category of all directed d-groups which are d-closed with d-homomorphisms and let $\boldsymbol{L}$ be the category of all l-groups with l-homomorphisms such that any l-group G is considered to be a d-group $(G, \oplus_m)$. Clearly, $\boldsymbol{L}$ is then a full subcategory in $\boldsymbol{D}$.

At first, we need to show that for any $(G, \oplus) \in \boldsymbol{D}$ the insertion

$$\alpha : (G, \oplus) \longrightarrow (L_d(G), \oplus_m)$$

is a morphism in $\boldsymbol{D}$, i.e. a d-homomorphism.

8.12. LEMMA. *For any* $(G, \oplus) \in \boldsymbol{D}$ *the canonical map* $\alpha : (G, \oplus) \longrightarrow (L_d(G), \oplus_m)$ *is a d-homomorphism.*

Proof. Let $x, y \in G_+$, $z \in x \oplus y$. We need to show that $\alpha(z) \wedge \alpha(x) = \alpha(z) \wedge \alpha(y) = \alpha(x) \wedge \alpha(y)$ in $L_d(G)$. Let

$$t \in \alpha(x) \wedge \alpha(y) = (x \cdot G_+ \cup y \cdot G_+)_{d_a}.$$

Then there exists a finitely-generated d-ideal Y_d such that $t \cdot Y_d \subseteq (xG_+ \cup yG_+)_d \times Y_d = ((xG_+ \cup yG) \cdot Y)_d$. Hence, for any $a \in Y_d$ there exist $y_1, \ldots, y_n \in Y$, $t_1, \ldots, t_n \in xG_+ \cup yG_+$ such that

$$at \in y_1 t_1 \oplus \ldots \oplus y_n t_n .$$

Let k be such that $t_1, \ldots, t_k \geqslant x$, $t_{k+1}, \ldots, t_n \geqslant y$. Then for $n \geqslant j \geqslant k+1$ we have

$$t_j \in (y)_d \subseteq (x\,,y)_d = (xG_+ \cup zG_+)_d$$

since $y \in x \oplus z$. Hence,

$$y_j t_j \in (xG_+ \cup zG_+) \cdot Y \subseteq (xG_+ \cup zG_+)_d \times Y_d = J$$

and for $1 \leqslant j \leqslant k$ we have

$$y_j t_j \in G_+ : Y \subseteq J\,.$$

Hence, $a \cdot t \in J$ and

$$t \cdot Y_d \subseteq (xG_+ \cup zG_+)_d \times Y_d,\quad t \in (xG_+ \cup zG_+)_d = \alpha(x) \wedge \alpha(z)\,.$$

The rest may be done similarly. ∎

In what follows we need to describe a functional character of the construction of a Lorenzen d-group (a similar construction may be done for any Lorenzen r-group).

So, let $\varphi : (G, \oplus) \longrightarrow (H, \oplus')$ be a morphism in D. Then $L_d(\varphi)$ is a morphism

$$L_d(\varphi) : L_d(G) \longrightarrow L_d(H)$$

such that

$$L_d(\varphi)\left(\frac{X_{d_a}}{Y_{d_a}}\right) = \frac{(\varphi(X))_{d_a}}{(\varphi(Y))_{d_a}}\,.$$

It is easy to see that this definition is correct. Moreover, we have

$$L_d(\varphi)\left(\frac{A_{d_a}}{C_{d_a}} \wedge \frac{B_{d_a}}{C_{d_a}}\right) = \frac{(\varphi(A \cup B))_{d_a}}{(\varphi(C))_{d_a}} = \frac{(\varphi(A) \cup \varphi(B))_{d_a}}{(\varphi(C))_{d_a}} =$$

$$= \frac{(\varphi(A))_{d_a}}{(\varphi(C))_{d_a}} \wedge \frac{(\varphi(B))_{d_a}}{(\varphi(C))_{d_a}} = L_d(\varphi)\left(\frac{A_{d_a}}{C_{d_a}}\right) \wedge L_d(\varphi)\left(\frac{B_{d_a}}{C_{d_a}}\right),$$

and $L_d(\varphi)$ is an l-homomorphism. Similarly, it may be proved that $L_d(\varphi)$ is a group homomorphism.

8.13. THEOREM. *The category* L *is a full reflective subcategory of* D.

P r o o f. It is easy to see that L_d is a functor. Moreover, according to 8.11, for $(G, \oplus) \in D$ the canonical map

$$\alpha_G = \alpha : (G, \oplus) \longrightarrow (L_d(G), \oplus_m)$$

is a d-homomorphism. We show that α is a reflection. To show it let $H \in L$ and let β be a d-homomorphism $(G, \oplus) \longrightarrow (H, \oplus_m)$. It is easy to see that for any l-group H we have $L_d(H) = H$ and that for any d-homomorphism $\varphi : G \longrightarrow G'$ we obtain a commutative diagram

$$\begin{array}{ccc} (G, \oplus) & \xrightarrow{\ \varphi\ } & (G', \oplus') \\ \Big\downarrow \alpha_G & & \Big\downarrow \alpha_{G'} \\ L_d(G) & \xrightarrow{\ L_d(\varphi)\ } & L_d(G') \end{array}$$

Hence, for a d-homomorphism β we have the following commutative diagram

$$\begin{array}{ccc} (G, \oplus) & \longrightarrow & L_d(G) \\ & {\scriptstyle\beta}\searrow \quad \swarrow{\scriptstyle L_d(\beta)} & \\ & (H, \oplus_m) & \end{array}$$

Finally, $L_d(\beta)$ is such that

$$L_d(\beta)\left(\frac{(x_1, \ldots, x_m)_{d_a}}{(y_1, \ldots, y_n)_{d_a}}\right) = (\beta(x_1) \wedge \ldots \wedge \beta(x_m)) \cdot (\beta(y_1) \wedge \ldots \wedge \beta(y_n))^{-1}$$

and it follows that $L_d(\beta)$ is unique which completes the above diagram. Therefore, the theorem is proved. ∎

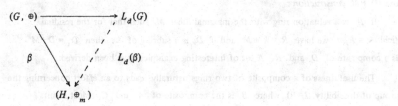

9. EXACT SEQUENCES OF GROUPS OF DIVISIBILITY.

M. Nagata [97] investigated an example of constructing the valuation ring using two other valuation rings in the following way.

Let C be a valuation ring and let A be a valuation ring of the field $k = C/M$ where M is the maximal ideal of C. Then if we denote by $\varphi : C \longrightarrow k$ the canonical map, $B = \varphi^{-1}(A)$ is a valuation ring called the *composite of* C *and* A.

This construction was generalized by J.Ohm [105] in the following way.

Let w be an additive semivaluation of a field K and let M_w be the maximal ideal of A_w. Let φ be the canonical homomorphism of A_w onto the residual field $k = A_w/M_w$. Let u be a semi-valuation of k and let ν be a semivaluation of K associated with a domain $A_\nu = \varphi^{-1}(A_u)$. Then ν is said to be *composite with* w *and* u, or a domain A_ν is said to be the *composite of* A_w *and* A_u.

Many examples in the literature can be interpreted as the composite of two rings, for example Prüfer [108] was the first who used such a construction. Other examples may be found in Krull [79], Gilmer, Parker [49], Henriksen [62]. One of the most useful constructions of such a type is the Gilmer, Heinzer's construction called the *$D + M$ construction*:

If R is a valuation ring with the maximal ideal M such that for the residual field $k = R/M$ we have $R = k + M$ and if D is a subring of k, then $D_1 = D + M$ is a composite of D and R. A lot of interesting examples have been derived.

The usefulness of a composite of two rings naturally leads to an effort concerning the group of divisibility $G(B)$ where B is the composite of A and C. Most results presented in this chapter are due to J.Ohm [105] and Mott, Schexnayder [95].

Let B be the composite of A and C, then we may consider the following exact sequence

$$(9.1)\quad 0 \longrightarrow U(C)/U(B) \xrightarrow{\alpha} K/U(B) \xrightarrow{\beta} K/U(C) \longrightarrow 0$$

where α is an embedding and $\beta(xU(B)) = xU(C)$.

9.2. THEOREM. *If we define the positive cone of* $U(C)/U(B)$ *as* $(U(C) \cap B)/U(B)$, *then* $U(C)/U(B) \cong_0 G(A)$ *and (9.1) is a lex-exact sequence*

$$(9.2)\quad 0 \longrightarrow G(A) \xrightarrow{\ \alpha\ } G(B) \xrightarrow{\ \beta\ } G(C) \longrightarrow 0 .$$

Proof. Let u, v, w be semivaluations with domains A, B, C, respectively. For $xU(B) \in U(C)/U(B)$ we set

$$\rho(xU(B)) = u(x) \in G(A)$$

Then clearly, ρ is the required o-isomorphism. It remains to show that (9.1) is lex-exact. But it follows directly using the fact, that

$$G(B)_+ = \{v(x) : x \in K,\ w(x) > 0 \text{ or there exists } y \in U(C) \cap B \text{ such that } v(x) = v(y)\}. \qquad \blacksquare$$

The following theorem enables us to construct a new one from two groups of divisibility.

9.3. THEOREM. *Let* J *be a group of divisibility of a quasilocal domain* C *with residual field* k *and let* H *be a group of divisibility of a subring* A *of* k. *If* J *is free (as a Z-modul), then* $H \oplus_L J$ *is a group of divisibility.*

Proof. Let B be the composite of A and C. By 9.2, there is a lex-exact sequence

$$0 \longrightarrow H \longrightarrow G(B) \longrightarrow J \longrightarrow 0 .$$

Since J is a free Z-modul, this sequence splits. In chapter 1 we have observed that it splits lexicographically and there exists an o-isomorphism ρ such that the following diagram commutates.

$$\begin{array}{ccccccccc}
0 & \longrightarrow & H & \longrightarrow & G(B) & \longrightarrow & J & \longrightarrow & 0 \\
 & & \uparrow{\scriptstyle =} & & \uparrow{\scriptstyle \rho} & & \uparrow{\scriptstyle =} & & \\
0 & \longrightarrow & H & \longrightarrow & H \oplus_L J & \longrightarrow & J & \longrightarrow & 0
\end{array}$$

Therefore, $H \oplus_L J$ is a group of divisibility. ∎

From the proof of 9.3 it follows that it is important to know whether a sequence (9.2') splits. The following proposition solves this problem completely.

9.4. **PROPOSITION.** *The sequence (9.2') splits if and only if there exists a set* $M = \{x_c : x_c \in K,\ c \in G(C)\}$, *such that* $w(x_c) = c$ *and* $(x_c \cdot x_d)/x_{c+d} \in U(B)$.

P r o o f. Let M satisfy the conditions. Then the set $\{\nu(x_c) : x_c \in M\}$ forms a subgroup G of $G(B)$ isomorphic to $G(C)$, and $G(C)$ is isomorphic to $\alpha(G(A)) \oplus G$.

Conversely, if the sequence splits, then there exists a subgroup S of $G(B)$ such that β maps isomorphically onto $G(C)$. Let b_c, $c \in G(C)$, denote the element of S such that $\beta(b_c) = c$. Then $b_c + b_d = b_{c+d}$. Let choose $x_c \in K$ such that $\nu(x_c) = = b_c$. Then $M = \{x_c : c \in G(C)\}$ has the required properties. ∎

9.5. **REMARKS.** (1) Let J be an arbitrary o-group. Following the results of chapter 8 on the quotient field K of the group algebra $k[J]$ of a group J over a field k, there is a valuation w such that for $X^g \in k[J]$, $w(X^g) = g$ holds. Let H be a group of divisibility of a domain A in a field k and let G be a group of divisibility of a composite of A and R_w. Then the exact sequence

$$0 \longrightarrow H \longrightarrow G \longrightarrow J \longrightarrow 0$$

satisfies the conditions of 9.4, hence, it splits and it follows that it splits lexicographically. Therefore, $G \cong_0 H \oplus_L J$ and we obtain the following result:

If H is a group of divisibility and J is an o-group, then $H \oplus_L J$ is a group of divisibility.

(2) In the construction mentioned above the domain B was the direct sum of A and the maximal ideal M of R_w. J. L. Mott [91] posed the following question: If B is the composite of C and A such that B is the direct sum of A and the maximal ideal M of C, does the sequence (9.2') necessarily split? This question has a negative answer, as shown by J.L.Mott [91] in the following way.

Suppose the sequence

$$(+)\quad 0 \longrightarrow G \longrightarrow H \xrightarrow{\beta} J \longrightarrow 0$$

is lex-exact and let H be an l-group. By 8.1, there is a Bezout domain B such that $G(B) = H$. It is easy to see that if (+) is lex-exact and H is an l-group, then J is an o-group and G is an l-ideal of H. Moreover, each filter of H_+ under containment compares with the prime filter $H_+ - G$. By Yakabe's bijection (chapter 2) there is a prime ideal P of B such that $w_A(P - \{0\}) = H_+ - G$ and $G(B_P) \cong_0 J$ by 2.3. Thus, B_P is a valuation ring. Since every filter of H_+ compares with $H_+ - G$, it follows that every ideal of B compares with P. Thus, B is the composite of B_P and B/P, and it is possible to show that B is the direct sum of P and B/P.

Now, suppose that $\mathrm{Ext}(J, G) \neq 0$ where J and G are torsion-free groups and suppose, moreover, that H is a nonsplitting extension of G by J. Then, let G and J be totally ordered, and let H be totally ordered by

$$H_+ = \{h \in H : h \in G_+ \text{ or } \beta(h) > 0\}.$$

The sequence (+) is lex-exact and the rings B, B_P, B/P answer Mott's question in the negative.

In capters 2 and 3 we considered several methods for constructing po-groups which are not groups of divisibility. J. Ohm [105] using exact sequences of po-groups created another such method.

In a sequel we say that an l-group G satisfies the condition (+) if there exists an l-realization $\rho : G \longrightarrow \prod_{i \in I} G_i$ such that there are $g_1, g_2 \in G$, $g_1 \parallel g_2$, such that $\rho_i(g_1) \neq \rho_i(g_2)$ for every $i \in I$.

9.6. **THEOREM.** *Let G be an l-group satisfying (+) and let* $0 \longrightarrow H \xrightarrow{\alpha} J \xrightarrow{\beta} G \longrightarrow 0$ *be a lex-exact sequence, $H \neq \{0\}$. Then J is not a group of divisibility.*

We prove at first the following proposition.

9.7. **PROPOSITION.** *Let* $0 \longrightarrow A \xrightarrow{\alpha} B \xrightarrow{\beta} C \longrightarrow 0$ *be a lex-exact sequence of po-groups. If A is directed, then β is a v-homomorphism. If A is not directed, then β is a v-homomorphism if and only if C satisfies the following:*

(9.8) *for* $c_1, c_2 \in C$, $c_2 > c$ *for all* $c < c_1$ *imply* $c_2 \geq c_1$.

P r o o f. Let $b, b_1, \dots, b_n$ be elements of B such that $b \in \sup(\inf_B(b_1, \dots, b_n))$. Let $d' \in C$ be such that $d' \leqq \beta(b_1), \dots, \beta(b_n)$ and let $d \in B$ be such that $\beta(d) = d'$, then $\beta(b_i - d) \geqq 0$.

Consider first the case when A is directed. If $\beta(b_i - d) > 0$ then $b_i - d \in \alpha(A)$. In either case, $d \leqq b_i - a_i$ for some $a_i \in \alpha(A)$. Since A is directed, there exists $a \in \alpha(A)$ such that $a \leqq a_1, \dots, a_n$. Therefore, $d \leqq b_i - a$ and, hence $d + a \leqq$ $\leqq b_1, \dots, b_n$. Then $b \in \sup(\inf_B(b_1, \dots, b_n))$ implies $b \geqq d + a$. Thus, $\beta(b) \geqq$ $\geqq \beta(d) = d'$ which proves the first assertion.

Now, consider the case that A is not directed and C satisfies (9.8). Suppose that c is an element of C such that $c < d'$. For any $c' \in B$ such that $\beta(c') = c$, $\beta(b_i) \geqq d' > c$ implies $b_i \geqq c'$. Thus, $b \geqq c'$. Since this holds for any pre-image of c, we also conclude that $b \geqq c + a$ for any $a \in \alpha(A)$. If $b = c'$, then $0 \geqq a$ for any $a \in \alpha(A)$, which would imply that $A = \{0\}$ on the contrary to the assumption that A is not directed. Thus, $b > c'$ for any $c' \in B$ such that $\beta(c') = c$. This implies that $\beta(b) > \beta(c') = c$. By (9.8), $\beta(b) \geqq d'$.

Finally, we assume that β is a ν-homomorphism and let c_1, c_2 be elements of C such that $c_2 > c$ for all $c < c_1$. Choose a pre-image c_1' of c_1 in B. Since A is not directed, we can find $a_1, a_2 \in \alpha(A)$ such that there does not exist $a \in \alpha(A)$ with $a \leqq a_1, a_2$. Suppose that $f \leqq c_1' + a_1, c_1 + a_2$. Then $\beta(f) \leqq c_1$. If $\beta(f) =$ $= c_1$, then $f = c_1' + a$ for some $a \in \alpha(A)$, $a \leqq a_1, a_2$, a contradiction. Thus, $\beta(f) < c_1$. Therefore, by the initial assumption, $c_2 > \beta(f)$. Hence, for any pre-image $c_2' \in B$ of c_2, $c_2' \geqq f$. By the choice of f and the assumption that β is a ν-homomorphism, we conclude that $c_2 \geqq \inf(\beta(c_1' + a_1), \beta(c_1' + a_2)) = c_1$. ∎

9.9. COROLLARY. *If* $0 \longrightarrow A \xrightarrow{\alpha} B \xrightarrow{\beta} C \longrightarrow 0$ *is lex-exact and* C *is an l-group, then* β *is a* ν*-homomorphism.*

P r o o f. Any *l*-group satisfies (9.8).

9.10. LEMMA. *Suppose that* $0 \longrightarrow A \xrightarrow{\alpha} B \xrightarrow{\beta} C \longrightarrow 0$ *is lex-exact and* v *is a map of* K^* *onto* B *(*K *is a field) and let* $w = \beta \cdot v$. *If* $A \neq \{0\}$ *and if there exist* $x, y \in K^*$ *such that* $w(x+y) < w(x), w(y)$, *then* v *is not a semi-valuation.*

P r o o f. $w(x+y) < w(x), w(y)$ implies $v(x+y) + a \leqq v(x), v(y)$ for all

$a \in \alpha(A)$. Therefore, if ν is a semivaluation, we must have $\nu(x+y) \geqq \nu(x+y)+a$ for all $a \in \alpha(A)$. However, this implies $A = \{0\}$, a contradiction. ∎

9.11. **LEMMA.** *Suppose that* C *is an l-group and let* $\rho : C \longrightarrow \prod_{i \in I} C_i$ *be an l-realization of* C. *Suppose that* w *is a semivaluation of a field* K *with a group* C *and let* C *satisfy* (+). *Then there exist* $x_1, x_2 \in K^*$ *such that* $w(x_1 + x_2) < w(x_1)$, $w(x_2)$.

Proof. Choose $x_1, x_2 \in K^*$ such that $w(x_i) = c_i$ and let $w_i = \rho_i \cdot w$ be a valuation of K. Since $w_i(x_1) \neq w_i(x_2)$ we have $w_i(x_1 + x_2) = \min(w_i(x_1), w_i(x_2))$. Then, since $\rho w = \prod_{i \in I} w_i$, $w(x_1 + x_2) = \inf(w(x_1), w(x_2))$ too. However, $\inf(w(x_1), w(x_2)) < w(x_1)$, $w(x_2)$ according to our hypothesis that c_1 and c_2 are unrelated. ∎

We may return to the proof of 9.6. Suppose that B is a group of divisibility and let ν be a semivaluation of a field K possessing the group B. By 9.9, β is a ν-homomorphism. Therefore, $w = \beta \cdot \nu$ is a semivaluation, and by 9.11., there exist $x_1, x_2 \in K^*$ such that $w(x_1 + x_2) < w(x_1)$, $w(x_2)$, thus by 9.10, ν is not a semivaluation.

J.L. Mott and M. Schexnayder [95] have observed that if B is an integral domain in a field K and S is a saturated multiplicative system in A, the following exact sequence

$$0 \longrightarrow [w_B(S)] \longrightarrow G(B) \longrightarrow G(B_S) \longrightarrow 0$$

is o-exact where $[w_B(S)]$ is a subgroup of $G(B)$ generated by $w_B(S)$. The natural question arising here is whether $G(B) \cong_0 [w_B(S)] \oplus G(B_S)$.

More generally if

$$(9.12) \quad 0 \longrightarrow G \xrightarrow{\alpha} H \xrightarrow{\beta} J \longrightarrow 0$$

is an o-exact sequence, we say that this sequence *splits cardinally*, if $H \cong_0 G \oplus J$.

9.13. **LEMMA.** *The following conditions are equivalent for an o-exact sequence (9.12).*

(1) (9.12) splits cardinally.

(2) There is a spliting map $\gamma : H \longrightarrow G$ *such that*

i) γ *is an o-homomorphism,*

ii) for each $h \in H_+$, $\alpha\gamma(h) \leqq H$.

(3) There is a spliting map $\delta : J \longrightarrow H$ *such that*

i) δ *is an o-homomorphism,*

ii) for each $h \in H_+$, $\delta\beta(h) \leqq h$.

The proof is straightforward.

Now, if B is an integral domain and if $\{p_i\}_{i\in I}$ is a collection of prime elements such that (i) p_i and p_j are not associated for $i, j \in I$, $i \neq j$, (ii) $\bigcap_{n\in N} p_i^n B = (0)$ for each $i \in I$, and (iii) each $b \in B$ is divisible by only finitely many of the primes p_i, then we say that $\{p_i\}_{i\in I}$ has the *UF-property*.

9.14. PROPOSITION. *Let* B *be an integral domain with the quotient field* K *and suppose that* $\{p_i\}_{i\in I}$ *is a collection of prime elements of* B *having the UF-property. Let* S *be a multiplicative system of* B *generated by* $\{p_i\}_{i\in I}$. *Then*

$$G(B) \cong {}_0 [w_B(S)] \oplus G(B_S),$$

and

$$[w_B(S)] \cong {}_0 \sum_{i\in I} Z_i, \quad Z_i = \mathbf{Z}, \quad i \in I.$$

Proof. Each $b \in B$ can be written as $b = s \cdot b'$ where $s \in S$, and b' is divisible by none of the primes p_i. Let $G = [w_B(S)]$, $H = G(B)$ and $J = G(B_S)$. Then we define $\gamma(w_B(b)) = w_B(s)$. The conclusion follows immediately from 9.13. ∎

Then we have the following

9.15. COROLLARY. *Let* B *and* S *be the same as in 9.14. If* B_S *is a GCD--domain (pseudo-principal, or UFD), then* B *is a GCD-domain (pseudo-principal, or UFD).*

Using 9.14 it is possible to improve the well-known result of Nagata [98].

9.16. COROLLARY. *Suppose that* B *is an integral domain such that each nonzero, nonunit of* B *is a finite product of irreducible elements and suppose that*

S is a multiplicative system generated by a collection of prime elements of B. If B_S is a UFD, then B is a UFD.

P r o o f. The primes in S satisfy the *UF*-property.

Now, let $B = A[X]$, where $X = \{X_i\}$ is a collection of indeterminates over A. It is easy to see that the *o*-exact sequence

(9.17) $0 \longrightarrow G(A) \longrightarrow G(A[X]) \longrightarrow G(k[X]) \longrightarrow 0$

always splits as a sequence of (abstract) groups since $G(k[X])$ is a free abelian group (k is the quotient field of A). But, a spliting map need not be an *o*-homomorphism. The next proposition shows that (9.17) splits cardinally if and only if A is a GCD--domain.

9.18. **PROPOSITION.** *Suppose that A is an integral domain with the quotient field k. Then the following are equivalent.*

(1) A is a GCD-domain.

(2) $G(A[X]) \cong_0 G(A) \oplus G(k[X])$.

(3) Each nonzero polynomial in B can be written (uniquely up to units) as a product of an element of A and a prime polynomial.

The proof may be found in Mott, Schexnayder [95].

10. GROUPS OF DIVISIBILITY OF KRULL DOMAINS.

One of the most intensively investigated divisibility groups is that of the Krull domain. A notion of a Krull domain (or discrete principal order by Krull) appeared in Krull's work in 1930 and considerably influenced the ring theory and the theory of numbers.

An integral domain A with the quotient field K is said to be a *Krull domain* provided there is a defining family $\{v_i\}_{i\in I}$ of discrete rank one valuations of K such that

(1) $A = \bigcap_{i\in I} R_{w_i}$,

(2) given $a \in A$, $a \neq 0$, there is at most a finite number of $i \in I$ such that $v_i(a) \neq 0$.

For properties and history of Krull domains we refer the reader to Nagata [97], Bourbaki [13], Gilmer [42], Fossum [37].

The following very important version of the approximation theorem holds for Krull domains.

10.1. THEOREM. *Let A be a Krull domain with the quotient field K and with a defining family $\{v_i\}_{i\in I}$. Let n_i be a given integer for each $i \in I$ such that $n_i = 0$ for almost all i. For any preassigned $i_1, \ldots, i_k \in I$ there is an $x \in K$ such that $v_{i_t}(x) = n_{i_t}$, $t = 1, \ldots, k$; with $v_i(x) \geqq 0$ otherwise.*

Using the fact $G(R_v) \cong_0 \mathbf{Z}$ for a discrete rank one valuation v and after using 10.1, we may deduce then that a group of divisibility $G(A)$ of a Krull domain A is an order subgroup of an *o*-sum $\mathbf{Z}^{(I)}$ of card I copies of $\mathbf{Z}$. Such groups in Jaffard's [65] terminology are called *normal*. We note, however, that one ought to make a distinction between a normal group and a group of divisibility G of a Krull domain since the last admits the following approximation property that is a simple transcription of 10.1 :

To any element $\alpha \in G \subseteq \mathbf{Z}^{(I)}$ and any finite subset $J \subseteq I$ there exists an element $\delta \in G$ such that $\alpha(j) = \delta(j)$ for all $j \in J$ and $\delta(i) \geqq \alpha(i)$ otherwise.

The natural question arising here is whether each normal group with the above approximation property is a group of divisibility (necessary of a Krull domain). The answer is negative, but the complete description of such a group is not known, although L. Claborn (see Fossum [37] has obtained considerable results in this direction. We show the principal Claborn's results in a sequel.

A step towards a simplification of a notion of a Krull domain is taken by Borevic-Shafarevic [12], where the notion of a theory of divisors of a domain A is introduced as a map φ from the positive cone of a group of divisibility G of A into a free semigroup D ordered by the division relation verifying the following three conditions:

(1) φ is an o-isomorphism of G_+ into D,

(2) if $\varphi(w_A(a)) \geqq \alpha$ and $\varphi(w_A(b)) \geqq \alpha$, $a \pm b \neq 0$, then $\varphi(w_A(a \pm b)) \geqq \alpha$, $\alpha \in D$,

(3) if α, β are elements of D such that

$$\{g \in G : \varphi(g) \geqq \alpha\} = \{g \in G : \varphi(g) \geqq \beta\}$$

then $\alpha = \beta$.

The elements of D are called divisors. Using the approximation theorem 10.1 it is possible to show the propositions 'to be a Krull domain' and 'to be a domain with a theory of divisors' are, in fact, equivalent. For example, if A is a Krull domain with the defining family $\{w_i\}_{i \in I}$, we denote by D the free semigroup ordered by division with the set I as the set of free generators. For every $x \in A$ we define a map $\varphi : G(A)_+ \longrightarrow D$ in the following way:

$$(\varphi(w_A(x)))(i) = w_i(x) \in Z, \quad i \in I,$$

where we identify elements of D with sequences indexed by I. Then it is possible to show that φ is a theory of divisors of A.

It is very interesting and important that the above definition of a theory of divisors of a domain is not so explicitly based on the presence of the additive operation as follows from axiom (2), namely, it has been observed by L. Skula [121], that axiom (2) is redundant. This fact enables us to define a theory of divisors for a po-group G without the assumption that G is a group of divisibility.

A systematical approach of this has been done by Skula [121], [122], [123], the first purely multiplicative treatment in this direction is Clifford's paper [24].

Let us now elucidate the principal Skula's results.

10.2. DEFINITION. Let G be a directed *po*-group. Then an *o*-isomorphism h of G into a directed *po*-group D such that D_+ is a free semigroup (in this case D is an *l*-group) is called a *theory of divisors* if

(i) $\forall \alpha, \beta \in D_+$, $\{g \in G : h(g) \geqq \alpha\} = \{g \in G : h(g) \geqq \beta\}$ implies $\alpha = \beta$.

In this case we say that G is a *δ-group.*

It should be observed that if $h : G \longrightarrow D$ in an *o*-isomorphism of a directed *po*-group G into a directed *po*-group D such that D_+ is free (such a *po*-group will be called a *UF-group*), then the condition (*i*) is equivalent to the following one:

(ii) $\forall \alpha \in D_+ \; \exists g_1, \ldots, g_n \in G$ such that $\alpha = h(g_1) \wedge \ldots \wedge h(g_n)$

If for an *o*-isomorphism h considered before the following condition (iii) holds instead of (ii), then G is called a *δ_1-group.*

(iii) $\forall \alpha, \beta \in D_+ \; \exists \gamma \in D_+$ such that $\beta \wedge \gamma = 1$, $\alpha \cdot \gamma \in h(G)$.

It is possible to show that every δ_1-group is a δ-group and, moreover, if G is a group of divisibility of a domain A with the theory of divisors $\varphi : G_+ \longrightarrow D$, then φ may be extended onto a map from G, and in this case G is a δ_1-group (see Skula [121]). Hence, δ_1-groups constitute an overclass of the class of divisibility groups of Krull domains.

Further, a theory of divisors of a *po*-group G is unique (if it exists) in the following sense. If $h_1 : G \longrightarrow D_1$, $h_2 : G \longrightarrow D_2$ are the theory of divisors of G, then there exists an isomorphism f such that the following diagram commutates.

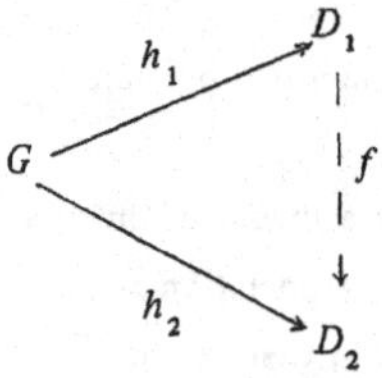

Thus, if G is a δ-group, we sometimes denote by cG a *UF*-group for which there exists a homomorphism $h : G \longrightarrow cG$ creating a theory of divisors.

The natural questions arising here are the following two ;

(A) Is it possible to describe all δ-groups (δ_1-groups) ?

(B) Is every δ_1-group a group of divisibility ?

Question (A) was completely solved by Skula [121], but the answer of (B) is, unfortunately, negative as was shown by Claborn who, moreover, gave a partial positive solution of (B). We now describe the principal ideas of their solutions.

At first, to solve (A) we need an analogue of a class group of a Krull domain constructed for a δ-group. Let us recall that for a Krull domain A the class group $C(A)$ of A is the factor group of a group of fractional v-ideals of A according to the group of principal v-ideals. If we realize that for a theory of divisors $\varphi : G(A) \longrightarrow$ $\longrightarrow D$ of a Krull domain A the *UF*-group D is isomorphic with a group of v-ideals and φ is an insertion $x \longmapsto (x)_v$, the class group $\mathcal{C}(A)$ is the factor group $D/\sim$ where

$$\alpha \sim \beta \text{ iff } \exists g_1, g_2 \in G(A)_+, \ \varphi(g_1) \cdot \alpha = \varphi(g_2) \cdot \beta .$$

This leads us to the following construction. Let $h : G \longrightarrow D$ be a theory of divisors. For $\alpha, \beta \in D$ we set

$$\alpha \sim \beta \text{ iff } \exists g_1, g_2 \in G, \ h(g_1) \cdot \alpha = h(g_2) \cdot \beta .$$

Then $\sim$ is a congruence relation on D, and the factor group $\Gamma_h = D/\sim$ is called a *divisor class group of* h. We denote by $\varphi_h : D \longrightarrow \Gamma_h$ the canonical homomorphism.

10.3. DEFINITION. Let Γ be an abelian group. A set $M \subseteq \Gamma$ is called a *strong system of generators* for Γ if for every $g \in \Gamma$, $g \neq 1$, there exist $g_1, \ldots, g_k \in M$ and natural numbers $m_1, \ldots, m_k$ such that

$$g = g_1^{m_1} \ldots g_k^{m_k} .$$

10.4. PROPOSITION. *Let* $h : G \longrightarrow D$ *be a theory of divisors of* G *and let* $P(D)$ $(= P)$ *be the set of free generators of* D_+.

(1) For each $p \in P$ *the set* $\varphi_h(P - \{p\})$ *is a strong system of generators for* Γ_h.

(2) G *is a δ-group if and only if for every* $p_1, \ldots, p_n \in P$ *the set*

$\varphi_h(P - \{p_1, \ldots, p_n\})$ *is a strong system of generators for* Γ_h.

For the proof see Skula [121]. Then the following theorem completely solves question (A).

10.5. THEOREM. *Let* D *be an UF-group with a set* P *of free generators of* D_+ *and let* $\sim$ *be a congruence relation on* D *such that for canonical map* $\varphi : D \longrightarrow D/\sim$ *the following holds:*

$\forall p \in P$, $\varphi(P - \{p\})$ *is a strong system of generators of* $D/\sim$. *Then* $G =$ $= \ker \varphi$ *is the unique subgroup of* D *such that the embedding* $h : G \longrightarrow D$ *is a theory of divisors of* G. *If* $\Gamma = D/\sim$, *then* $\sim = \sim_h$, $\Gamma = \Gamma_h$, $\varphi = \varphi_h$.

Proof. At first, let $h(g_1) \leqq h(g_2)$ in D. Then there exists $\alpha \in D_+$ such that $g_1 . \alpha = g_2$. Then $\varphi(g_1) . \varphi(\alpha) = \varphi(g_2)$ and $\varphi(\alpha) = 1$, $\alpha \in G$. Thus, $g_1 \leqq g_2$ in G and h is an o-isomorphism into D. Now, let $\alpha_1, \alpha_2 \in D_+$, $\alpha_1 \neq \alpha_2$. We may assume that there exists $p \in P$ such that $p^n \leqq \alpha_2$, $p^n \nleqq \alpha_1$ for some natural n. Further, we may assume that $\alpha_1 \notin h(G)$. By the assumption we may find $p_1, \ldots,$ $p_k \in P - \{p\}$ and natural numbers $n_1, \ldots, n_k$ such that $\varphi(\alpha) = \varphi(p_1)^{n_1} \ldots$ $\varphi(p_k)^{n_k}$. If we set $\alpha = p_1^{n_1} \ldots p_k^{n_k}$, we have $\alpha \cdot \alpha_1 \in h(G)$ and $\alpha_2 \nleqq \alpha_1 \cdot \alpha$. Thus, h is a theory of divisors. Further, let $\alpha_1, \alpha_2 \in D$. If $\alpha_1 \sim_h \alpha_2$, then clearly $\alpha_1 \sim \alpha_2$. Conversely. let $\alpha_1 \sim \alpha_2$. Then $\varphi(\alpha_1) = \varphi(\alpha_2)$ and there exist $\alpha \in D$ such that $\varphi(\alpha) \cdot \varphi(\alpha_1) = \varphi(\alpha) \cdot \varphi(\alpha_2) = 1$. We set $g_1 = \alpha_2 \cdot \alpha$, $g_2 = \alpha_1 \cdot \alpha$. Then $g_1, g_2 \in$ $\in G$, and $g_1 \cdot \alpha_1 = g_2 \cdot \alpha_2$. Thus, $\sim = \sim_h$. The rest is clear. ∎

Now, let D be a *UF*-group and let Γ be an abelian group. We consider a map $\varphi : P(D) \longrightarrow \Gamma$ such that $\varphi(P(D) - \{p\})$ $(\varphi(P(D) - \{p_1, \ldots, p_k\}))$ is a strong system of generators for Γ for each $p \in P(D)$ (for each natural k and each $p_1, \ldots,$ $p_k \in P(D)$). Then the map φ may be extended onto an homomorphism of D onto Γ. If we set $G = \ker \varphi$, and denote by h the inclusion $G \hookrightarrow D$, it follows by 10.5 that h is a theory of divisors such that G is a δ-group (δ_1-group) with the divisor class group Γ_h isomorphic Γ.

Moreover, this construction fully describes the family of subgroups of D for which the inclusion creates a theory of divisors.

10.6. EXAMPLES. (1) Let p_1, p_2 be two different elements and let D be the *UF*-group such that $P(D) = \{p_1, p_2\}$. Let $\Gamma = \{\overline{0}, \overline{1}\}$ be the additive group $Z/(2)$. If we set $\varphi(p_1) = \varphi(p_2) = \overline{1}$, then $\varphi(P(D) - \{p\})$ is a strong system of generators for each p. By 10.5, the group

$$G = \ker \varphi = \{p_1^{n_1} \cdot p_2^{n_2} : n_1, n_2 \in Z,\ n_1 + n_2 \equiv 0 \bmod 2\}$$

is a δ-group.

(2) Let p_1, p_2, p_3, p_4 be different elements and let D be the *UF*-group such that $P(D) = \{p_1, p_2, p_3, p_4\}$. Let $\Gamma = (Z, +)$. We set $\varphi(p_1) = \varphi(p_2) = 1$, $\varphi(p_3) = \varphi(p_4) = -1$. Then $\varphi : P(D) \longrightarrow \Gamma$ is a map such that $\varphi(P(D) - \{p\})$ is a strong system of generators of Γ for each p. By 10.5., the group

$$G = \ker \varphi = \{p_1^{n_1} \cdot p_2^{n_2} \cdot p_3^{n_3} \cdot p_4^{n_4} : n_i \in Z,\ n_1 + n_2 = n_3 + n_4\}$$

is a δ-group.

(3) Let p_i, $i \in N$ $(= Z_+)$, be different symbols and let D be a *UF*-group with $P(D) = \{p_i : i \in N\}$. We set $\Gamma = (Z, +)$ and let

$$\varphi(p_i) = (-1)^i .$$

Then $\varphi : P(D) \longrightarrow \Gamma$ is a map such that $\varphi(P(D) - \{p_1, \ldots, p_k\})$ is a strong system of generators for each k and each p_i and by 10.5, the group

$$G = \ker \varphi = \{p_{i_1}^{n_1} \ldots p_{i_s}^{n_s} : s \in N,\ n_t \in Z,\ \sum_{t=1}^{s} n_t(-1)^{i_t} = 0\}$$

is a δ_1-group.

Clearly, every abelian group is a divisor class group for a theory of divisors of a *po*-group. In fact, let Γ be an abelian group, and let D be a *UF*-group such that $P(D) = \Gamma$. We let

$$\varphi : P(D) \longrightarrow \Gamma$$

be the identity map, then clearly $\varphi(P(D) - \{p\})$ is a strong system of generators for each $p \in \Gamma$. By 10.5, there is a theory of divisors h such that $\Gamma = \Gamma_h$.

K.E. Aubert [2] has investigated an interesting generalization of a theory of divisors which gives a common back ground for the investigation of Prüfer and Krull domains. This generalization depends on the fact that an *UF*-group D from the definition of a theory of divisors is replaced by an l-group. The exact definition is given in the following.

Let G be a directed *po*-group. Then an o-isomorphism h of G into an l-group D is called a theory of quasidivisors for G if

(i)' $\forall \alpha, \beta \in D_+$, $\{g \in G : h(g) \geqslant \alpha\} = \{g \in G : h(g) \geqslant \beta\}$ implies $\alpha = \beta$.

Clearly any theory of divisors is a theory of quasidivisors.

The existence of such a theory of quasidivisors for *po*-groups follows from the following theorem (Aubert [2]), proof of which may be essentially found in Jaffard [65].

10. 7. THEOREM. *A t-closed po-group* G *has a unique theory of quasidivisors determined by the canonical injection* G *into its Lorenzen t-group* $L_t(G)$.

Now, if A is a Prüfer domain, then a *po*-group $G = (G(A), \oplus_A)$ is d-closed and it follows that G is t-closed. Hence, G admits a unique theory of quasidivisors and in this case $L_t(G)$ is isomorphic to the group of all finitely generated t-ideals of $G(A)$. Moreover, if A is a Krull domain then again G is t-closed and it is possible to show that $L_t(G)$ is a *UF*-group; hence the theory of quasi-divisors $G \hookrightarrow L_t(G)$ is a theory of divisors.

A much more complicated situation emerges when we consider an analogous question dealing with the theory of divisors of a domain. An affirmative answer was obtained by L. Claborn in his excellent work [20], where he states that every abelian group is a class group of a Dedekind domain.

Now we are leading to deal with question (B) which is considerably more complicated than question (A), and its answer is negative in general.

At first, we need some notation. For an *UF*-group D and a subgroup $G \subseteq D$ we say that G is *(finitely) dense* in D if for each $Y \subseteq P(D)$ such that card $Y <$ card $P(D)$ (card $Y < \aleph_0$) and each $\alpha \in D_+$ there exists a $g \in G_+$ such that $g(p) = \alpha(p)$, $p \in Y$.

For every finitely dense subgroup G of a *UF*-group D, the inclusion $h : G \longrightarrow D$ is a theory of divisors, and G is a δ_1-group. In fact, according to 10.5, it suffices to show that for the canonical map $\varphi : D \longrightarrow \Gamma = D/G$, a set $\varphi(P(D)) - \{p_1, \ldots, p_k\})$ is a strong system of generators for each $p_1, \ldots, p_k \in P(D)$, i.e. for each $p_1, \ldots, p_k$ and each i, $1 \leqq i \leqq k$, it is necessary to find elements $g_1, g_2 \in G_+$ such that

$$(g_2 \cdot p_i)(p) \geqq g_1(p), \quad p \in P(D),$$
$$(g_2 \cdot p_i)(p) = g_1(p) \quad \text{for } p = p_1, \ldots, p_k.$$

Since G is finitely dense in D, we may find an element $g_1 \in G_+$ such that $g_1(p_i) = 1$, $g_1(p_j) = 0$, $j \neq i$, $j = 1, \ldots, k$. Analogously, there exists an element $g_2 \in G_+$ such that

$$g_2(p) = g_1(p) \text{ if } g_1(p) > 0,$$
$$g_2(p_j) = 0, \; j = 1, \ldots, k.$$

Then clearly g_1 and g_2 satisfy the required conditions.

10.8. THEOREM (Claborn). *For every UF-group D such that $m = \text{card}(P(D)) \geqq 2^c$ there exists a finitely dense subgroup G of D such that G is not a group of divisibility.*

P r o o f. Let $P(D)$ be partitioned into m subsets P_p, $p \in P(D)$, having a countable numbers of elements,

$$P(D) = \bigcup_{p \in P(D)} P_p, \quad \text{card } P_p = \aleph_0.$$

Let

$$G = \{\alpha \in D : \sum_{q \in P_p} \alpha(q) \text{ is even for all } p \in P(D)\}.$$

Then clearly G is a finitely dense subgroup of D. Suppose that there is an integral domain A such that $G(A) = G$. Then A is a Krull domain. If we denote by $Z(A)$ the set of minimal prime ideals of A, $G(A)$ is a free group with the family $Z(A)$ of free generators. Hence, we may identify $Z(A)$ with $P(D)$. Let us choose c elements in A and let R be a subring in A generated by these elements. Let card $R = c$. Each of c nonzero elements of R lies in only a finite number of prime ideals of A. Thus, there are at least c prime ideals in $Z(A)$ which contain nonzero elements of R. Hence, there must be a set B of at least 2^c ideals in $Z(A)$ whose elements intersect R in (0).

Let $p \in B$. Since every minimal prime ideal of a Krull domain may be generated by two elements, using the approximation theorem we may find two elements f and g such that $p = (f, g)$ with $w_p(f) = 1$, $w_p(g) = 2$. Consider the c elements $f + rg$, $r \in R$. Then $w_p(f + rg) = 1$, $r \in R$. Now, $w_A(f+rg) \in$ $\in G$, and it follows that $\sum_{q \in P_{q'}} w_q(f + rg)$ is even for $P_{q'}$ such that $p \in P_{q'}$. Hence, for every $r \in R$ we may find an element $q_r \in P_{q'} \cap B$ such that $q_r \neq p$, $w_{q_r}(f+rg) >$ > 0. Since card $R >$ card $(P_{q'} \cap B)$, there are $r' \neq r \in R$ such that $q = q_{r'} = q_r$. Hence, $f + rg$, $f + r'g \in q$ and it follows $(r - r')g \in q$. Since $q \in B$, we have $g \in q$ and it follows $f \in q$. Thus, $q = p$, a contradiction. ∎

So far the best result which describes *po*-groups being groups of divisibility of a Krull domain is the following theorem due to L. Claborn.

10.9. THEOREM (Claborn's Realization Theorem). *Let D be a UF-group and let G be a dense ordered subgroup of D. Then there is a Dedekind domain A and an o-isomorphism $\rho : cG(A) \longrightarrow D$ such that $\rho(G(A)) = G$.*

For the rather complicated proof see Fossum [37].

Using this theorem it is possible to construct examples of divisibility groups of Dedekind domains, for example, the group G from 10.6 (3) is dense in D.

Moreover, L. Skula [124] has used the Claborn's result to derive the existence of a Dedekind domain with a prescribed property. Namely, for a Dedekind domain A with the theory of divisors $h : G(A) \longrightarrow D$ he said that a pair $[\Gamma_h , \varphi(P(D))]$ is a *c-characteristic* of A. Then the following holds.

10.10. THEOREM. *Let M be a strong system of generators of a group Γ. Then there exists a Dedekind domain whose c-characteristic is $[\Gamma, M]$.*

P r o o f. (Sketch). Let $M \neq \phi$ be a strong system of generators of Γ. Let $\mathfrak{m}$ be an infinite cardinal number such that $\mathfrak{m} >$ card M, and for $\mu \in M$ let X_μ denote an arbitrary set with the property that card $X_\mu = \mathfrak{m}$ and for $\alpha, \beta \in M$, $\alpha \neq \beta$, $X_\alpha \cap X_\beta = \phi$ holds. Let D be a *UF*-group such that $P(D) = \bigcup_{\mu \in M} X_\mu$. For $p \in X_\mu$ we put $\varphi(p) = \mu$. Then φ may be extended onto a homomorphism $\varphi : D \longrightarrow \Gamma$.

We put $G = \ker \varphi$. Then it is possible to show that G is a dense subgroup of D, and by 10.9, there is a Dedekind domain A and an o-isomorphism ρ from $cG(A)$ such that $\rho(G(A)) = G$. Then the rest may be done using the basic properties of a divisor theory.

11. TOPOLOGICAL GROUPS OF DIVISIBILITY.

By 8.1, every l-group G is a group of divisibility of a Bezout domain. But, every l-group may be endowed with the discrete topology and, therefore, considered to be a topological l-group with continuous operations. We have an analogous situation for fields: every field may be considered to be a topological field with respect to the discrete topology. Hence, it seems natural to consider the following question. Does there exist for any topological group G a topological field K with a topology T and a domain A of K such that K is the quotient field of A, the factor topological group $K^*/U(A)$ is topologically o-isomorphic (i.e. o-isomorphic and homeomorphic) with G? In this case we say that G has a representation (K, T, A) and write $G = (K, T, A)$.

By a *topological l-group* (notation : *tl-group*) we shall mean a triple $(G, \leqq, F)$ where $(G, \leqq)$ is an l-group, (G, F) is a topological group and $(|G|, \leqq, F)$ is a topological lattice.

Two *tl*-groups are *tl-isomorphic* if there is a homeomorphism between them which is an l-isomorphism.

A set $\{H_i : i \in I\}$ of prime l-ideals of a *tl*-group G is a *t-realizator* of G if H_i is closed in G for every $i \in I$ and $\cap\{H_i : i \in I\} = 0$, and the natural map

$$\rho : G \longrightarrow \prod_{i \in I} G/H_i$$

is a *tl*-isomorphism from G onto $\rho(G)$ where $\rho(G)$ inherits its topology from the product $\Pi\, G/H_i$.

Further, for any field K and a valuation w of K with the value group G_w we may construct a field topology T_w of K defining the sets $U_{w,\alpha} = \{x \in K : w(x) > \alpha\}$, $\alpha \in G_w^+$, as a subbase of the neighbourhoods of zero in K. Then the group $U(R_w)$ is open in K^*, and $w : K^* \longrightarrow G_w$ is continuous with respect to the discrete topology on G_w.

As a first step towards the topological groups of divisibility we show a 'topological' version of 2.3. We use the following notation.

The symbol $S(A)$ $(S_0(A), S_c(A))$ denotes the set of all saturated multiplicative systems (open in a topological group K^*, closed in a topological group K^* respectively) of a domain A with the quotient field K where K is a topological field with a topology T; $O_0(G)$ $(O_c(G))$ denotes the set of all open (closed) c-ideals of a topological po-group G.

11.1. LEMMA. *Let G be a topological po-group and let H be a directed subgroup of G such that H_+ is closed in G. Then H is closed in G.*

Proof. Let $\overline{H}$ be the closure of H in G and let $\alpha \in \overline{H}$. Then for every neighbourhood U of zero in G there exists a neighbourhood V of zero such that $-V \subseteq U$. Since $(\alpha + V) \cap H \neq \phi$, there exists $\beta \in V$ such that $\alpha + \beta = \gamma \in H$. Since H is directed, we may find elements $\gamma_1, \gamma_2 \in H_+$ such that $\gamma = \gamma_1 - \gamma_2$. Thus, $-\gamma_1 + \alpha + \beta = -\gamma_2$, and $-\beta - \alpha + \gamma_1 = \gamma_2 \in H_+$, $-\beta + (-\alpha + \gamma_1) \in (U + + (-\alpha + \gamma_1)) \cap H_+$. It follows $-\alpha + \gamma_1 \in H_+$. Therefore, $-\alpha \in H_+ - \gamma_1 \subseteq H$ and H is closed in G. ∎

The proof of the following lemma is straightforward and is therefore omitted.

11.2. LEMMA. *Let A and B be domains with the quotient field K and for some subgroup H of $G(A)$ let there exist a group isomorphism ρ such that $w_B = \rho \cdot \varphi \cdot w_A$ where φ is the canonical map of $G(A)$ onto $G(A)/H$. Let $G(A) = (K, T, A)$, $G(B) = (K, T, B)$. Then $G(B)$ is homeomorphic with the factor topological group $G(A)/H$.*

11.3. PROPOSITION. *Let $G(A) = (K, T, A)$ and let $S \in S(A)$. Then $G(A_S) = (K, T, A_S)$ and $G(A_S)$ is homeomorphic with the factor topological group $G(A)/m(S)$.*

The proof follows directly from 11.2.

In the next theorem which is a topological version of Mott's bijection (see 2.3) we set $m_0 = m \mid S_0(A)$, $m_c = m \mid S_c(A)$ for a topological field K with a domain A.

11.4. **THEOREM.** *Let* $G(A) = (K, T, A)$. *Then* m_0 (m_c) *is a bijection between* $S_0(A)$ $(S_c(A))$ *and* $O_0(G(A))$ $(O_c(G(A))$ *if and only if* A^* *is open (closed) in* K^*. *If* A^* *is open in* K^*, *then* $S_0(A) = S(A)$, $O_C(G(A)) = O(G(A))$.

P r o o f. Let A^* be open in K^*. Then $U(A) = A^* \cap (A^*)^{-1}$ is open in K^*, and $G(A) = (K, T, A)$ is a discrete space, $O_0(G(A)) = O(G(A))$. Let $S \in S(A)$. Then since $m(S)$ is open in $G(A)$ and w_A is continuous, we obtain that $S = w_A^{-1}(m(S)) \cap A^*$ is open in A^*, hence $S \in S_0(A)$. Thus, $S(A) = S_0(A)$ and $m_0 = m$.

Conversely, let m_0 be a required bijection, then $m(A^*) = G(A) \in O_0(G(A))$ and it follows that A^* is open in K^*.

Let A^* be closed in K^*, $S \in S_c(A)$, $H = m(S)$, and let $\alpha = w_A(x) \in \overline{H}_+$ (the closure in $G(A)$). Let U be a neighbourhood of x in K^*, then there exists $z \in K^*$ such that $w_A(z) \in w_A(U) \cap H_+$, and for some $s \in S$, $a \in U$, $i, j \in U(A)$, we have $z = s \cdot j = a \cdot i$. Since S is saturated, we obtain $zi^{-1} \in U \cap S$ and $x \in \overline{S}$, the closure of S in K^*. Since S is closed, we have $\alpha \in H_+$, and H is closed by 11.1.

Further, let $H \in O_c(G(A))$, $S = m^{-1}(H)$. Since $w_A | A^*$ as a map from A^* into $G(A)_+$ is continuous, S is closed in A^* and S is then closed in K^*. Therefore, m_c is a bijection.

Conversely, if m_c is a bijection, from the fact $m(A^*) = G(A) \in O_c(G(A))$ it follows that A^* is closed in K^*. ∎

The following theorem solves completely the problem of existence of a representation for a topological *o*-group.

11.5. **THEOREM.** *Let* G *be a totally ordered tl-group. Then* G *has a representation if and only if* G *is a discrete space.*

P r o o f. If $G = (K, T, A)$, then a canonical map $w : K^* \longrightarrow G$ is a continuous valuation. Since every set $\{\beta \in G : \beta > \alpha\}$, $\alpha \in G_+$ is open in G (see Šmarda [132]), it follows that $T_w \leqslant T$. Thus, the set $U(R_w)$ is open in T and G is a discrete space. ∎

11.6. COROLLARY. *If a tl-group G has a representation, then every closed prime l-ideal of G is open in G.*

Let us observed that using 11.6, an example of a *tl*-group (non *o*-group) which has no representation is easy to construct. The topological product of two copies of an *o*-group with the interval non discrete topology works.

11.7. LEMMA. *Let (G, G) be a tl-group and let $G = (K, T, A)$. Then there exists a t-realizator of G if and only if there is a defining family of valuations $\{w_i : i \in I\}$ for A such that $T \geqq \sup(T_{w_i} : i \in I)$, and the set $\{U(R_{w_i}) : i \in I\}$ is a subbase for the sets $U \,.\, U(A)$, where U is an open neighbourhood of 1 in K^*.*

Proof. Suppose that $\{w_i : i \in I\}$ is a family of valuations of K that satisfies the conditions of 11.7. Then the value group G_i of w_i may be considered to be a *tl*-group with respect to the discrete topology. Then for every $i \in I$ there is a continuous and open *l*-epimorphism e_i which completes the following diagram

$$\begin{array}{ccc} K^* & \xrightarrow{\;w_A\;} & G(A) \\ {\scriptstyle w_i}\downarrow & & \downarrow{\scriptstyle \varphi} \\ G_i & \xleftarrow{\;e_i\;} & G \end{array}$$

where φ is a *tl*-isomorphism associated with the representation. Indeed, for every $\alpha \in G$ we set

$$e_i(\alpha) = w_i(x) \;, \quad \text{where } \varphi(w_A(x)) = \alpha \,.$$

Since $U(A) \subseteq U_i = U(R_{w_i})$, the definition is correct. Now, if $\alpha \geqq 0$ in G, for $x \in K$ such that $\varphi(w_A(x)) = \alpha$ we have $x \in A \subseteq R_{w_i}$ and $e_i(\alpha) = w_i(x) \geqq 0$. Conversely, since w_i is well centred on A, there exists $a \in A$ such that $w_i(a) = \alpha$ for any $\alpha \in G_i^+$. Further, since $T \geqq \sup T_{w_i}$, we obtain that $w_i^{-1}(U)$ is open for every $U \subseteq G_i$. Hence, $\varphi w_A(w_i^{-1}(U)) \subseteq e_i^{-1}(U)$ is open in G and e_i is open and continuous.

For every $i \in I$ there exists a closed (and open) prime *l*-ideal H_i of G such that the factor *tl*-group G/H_i is *tl*-isomorphic with G_i. We may identify these two groups.

Now, applying the identity $w_i = \epsilon_i \cdot \varphi \cdot w_A$, we obtain $H_i = \varphi w_A(U_i)$ and since $U(A) = \bigcap_{i \in I} U_i$, it follows that

$$\bigcap_{i \in I} H_i = \{0\}.$$

Finally, let $W \subseteq G$ be an open neighbourhood of zero in G. Then $\varphi^{-1}(W) = U \,.\, U(A)/U(A)$ for some open neighbourhood U of 1 in K^*. According to the assumption, there exist $i_1, \ldots, i_m \in I$ such that

$$U_{i_1} \cap \ldots \cap U_{i_m} \subseteq U \,.\, U(A)\,.$$

Hence, $H_{i_1} \cap \cdots \cap H_{i_m} \subseteq \varphi \cdot w_A(U_{i_1} \cap \cdots \cap U_{i_m}) \subseteq W$. Therefore, $\{H_i : i \in I\}$ is a subbase of the neighbourhoods of zero in G which meet at zero and according to Madell [77], this set is a *t*-realizator of G.

Conversely, suppose that $\{H_i : i \in I\}$ is a *t*-realizator of G and let w_i, $i \in I$, be the composition of the following maps:

$$K^* \xrightarrow{\ w_A\ } G(A) \xrightarrow{\ \varphi\ } G \xrightarrow{\ \pi\ } \pi(G) \subseteq \prod_{i \in I} G/H_i \xrightarrow{\ \epsilon_i\ } G/H_i$$

where π, ϵ_i are the canonical maps. Since φ, π, ϵ_i are *l*-homomorphisms, w_i is a valuation of K. It is clear that w_i is continuous and $\{w_i : i \in I\}$ is a defining family for A.

It is easy to see that for every $i \in I$ there exists isomomorphism κ_i which completes the following diagram:

$$\begin{array}{ccc}
G/H_i & \overset{\kappa_i}{- - - - - - -} & K^*/U_i = G_i \\
\uparrow {\scriptstyle \epsilon_i . \pi} & & \uparrow {\scriptstyle f_i} \\
G & \xrightarrow{\quad \varphi^{-1} \quad} & G(A)\,,
\end{array}$$

where f_i is the canonical map. Let $\kappa = \prod_{i \in I} \kappa_i$ and let us put $\kappa' = \kappa | \pi(G)$. Then κ' is a *tl*-isomorphism of $\pi(G)$ onto $H = \{(xU_i) \in \prod_{i \in I} G_i : x \in K^*\}$. Finally, we set

$$\rho = \kappa' \cdot \pi \cdot \varphi\,.$$

Then ρ is a homeomorphism of $G(A)$ onto H. Now, since w_i is continuous for every i, we have $T \geqq \sup T_{w_i}$. Let U be an open neighbourhood of 1 in K^*. Then $U \cdot U(A)/U(A)$ is open in $G(A)$, and $\rho(U \cdot U(A)/U(A))$ is open in H. Hence, there exist $i_1, \ldots, i_m \in I$ such that

$$\rho(\bigcap_{k=1}^{m} U_{i_k}/U(A)) = (\prod_{k=1}^{m} \{1\} \times \prod_{j \neq i_k} G_j) \cap H \subseteq \rho(U \cdot U(A)/U(A)).$$

Hence, $U_{i_1} \cap \cdots \cap U_{i_m} \subseteq U \cdot U(A)$ and this completes the proof. ∎

We say that a representation (K, T, A) of a *tl*-group G is *locally bounded* provided that (K, T) is a locally bounded topological field and $U(A)$ is a bounded set. Let us recall that a subset B of a topological field K is *bounded* if for every neighbourhood V of zero there exists another neighbourhood U with $B \,.\, U \subseteq V$. Then K is called *locally bounded* if there exists an open bounded set in K.

11.8. **THEOREM.** *Let G be a tl-group with a t-realizator $\{H_i : i \in I\}$. Then there exists a locally bounded representation of G if and only if I is a finite set and H_i is open for every $i \in I$.*

<u>Proof</u>. Suppose that I is a finite set and H_i is open for every $i \in I$. Then G is a discrete space. Let A be a Bezout domain with the quotient field K and such that $G(A) \cong_0 G$. Let $w_i = \epsilon_i \cdot \pi \cdot \varphi \cdot w_A$ be the same as in the proof of 11.7. We set $T = \sup T_{w_i}$. Then by Kowalsky [72], (K, T) is a locally bounded topological field. Since $\{w_i : i \in I\}$ is a defining family for A, $U(A)$ is an intersection of U_i and I is finite, it follows that $U(A)$ is open in K^* and $G = (K, T, A)$. It remains to prove that $U(A)$ is a bounded set. Let $W = U_{w_1}, \alpha_1 \cap \cdots \cap U_{w_m}, \alpha_m$, $\alpha_i \in G_i^+$. We may assume that valuations w_i are mutually independent. Using the approximation theorem for independent valuations, we may find an element $x \in K$ such that

$$w_i(x) > \alpha_i, \quad i = 1, \ldots, m.$$

Then $x \cdot U(A) \subseteq W$ and by Bourbaki [13], $U(A)$ is a bounded set.

Conversely, let (K, T, A) be a locally bounded representation of G. Thus, we may find a bounded neighbourhood U of zero in K, and, $((1 + U) \cap K^*) \cdot U(A)$

is a bounded set. By 11.7, there exists a defining family $\{w_i : i \in I\}$ for A such that $T \geqq \sup T_{w_i}$, and there exist $i_1, \ldots, i_m \in I$ such that

$$U_{i_1} \cap \cdots \cap U_{i_m} \subseteq ((1 + U) \cap K^*) \cdot U(A).$$

It follows that $(K, \sup T_{w_i})$ is a locally bounded topological field and, again by Kowalsky [72], I is finite and by 11.6, H_i is open for every $i \in I$. ∎

Further, we say that a representation $G = (K, T, A)$ is *locally compact* provided that (K, T) is a locally compact topological field.

11.9. THEOREM. *Let G be a tl-group with a t-realizator. Then there exists a locally compact representation of G if and only if G is a discrete space tl-isomorphic with Z.*

P r o o f. If (K, T, A) is a locally compact representation of G, then by 11.7, there exists a defining family $\{w_i : i \in I\}$ for A such that $T \geqq \sup T_{w_i}$. Every locally compact field is a complete topological field and it follows that T is a minimal field topology on K. Hence, $T = T_{w_i} = T_{w_j}$ and the valuations w_i, w_j, $i \neq j$, are mutually dependent. On the other hand, since T_{w_i} is locally compact, it follows that w_i is a discrete rank one valuation of K for every $i \in I$ and it follows that the valuations w_i are independent. Thus, card $I = 1$ and $H_1 = \{0\}$. Therefore, $G \cong_0 G/H_1 \cong_0 Z$. The converse is evident. ∎

As we have observed, there are *tl*-groups which have no representation. On the other hand, it is possible to construct examples of *tl*-groups (which are not *o*-groups) with a representation. For example, let us consider the group $Z^{(I)}$ of all integer valued sequences indexed by $I \subseteq N$ $(= Z_+)$ with a component wise operations. Then $Z^{(I)}$ is an *l*-group. Clearly, every such a group is a group of divisibility of a domain $A_I = \bigcap_{i \in I} R_{w_i} \subseteq Q$, where Q is the field of rationals and w_i is the p_i-adic valuation of Q for the ith prime number p_i. Let F be the topology on $Z^{(I)}$ with the subbase of neighbourhoods of zero consisting of prime ideals

$$H_i = \{\alpha \in Z^{(I)} : \alpha_i = 0\}, i \in I.$$

Then clearly $(Z^{(I)}, F)$ is a *tl*-group and if card $I = \aleph_0$, then F is a nondiscrete topology. The we have the following proposition.

11.10. PROPOSITION. $(Z^{(I)}, F) = (Q, \sup(T_{w_i} : i \in I), A_I)$.

<u>Proof</u>. At first we observe that $U(A)$ is closed in Q^*, since $U_i = U(R_{w_i})$ is closed for every $i \in I$. Let

$$\rho : G(A_I) = Q^*/U(A_I) \longrightarrow Z^{(I)}$$

be defined such that $\rho(w(x))(i) = \rho(xU(A_I))(i) = w_i(x)$, $i \in I$, where w is a semi--valuation associated with A_I. Clearly, ρ is an o-isomorphism. By Močkoř [84], to prove the proposition it remains to show that ρ is open and continuous. We have

$$\rho^{-1}(H_i) = w(U_i) = U_i/U(A_I),$$

and this set is an open neighbourhood of zero in $G(A_I)$ since $U_i = w_i^{-1}(0)$ is open in (Q, T) for $T = \sup T_{w_i}$. On the other hand,

$$\rho(U_{w_i,a}/U(A_I)) = \{\alpha \in Z^{(I)} : \alpha_i > a\} \ (= B)$$

as follows using the approximation theorem for independent valuations of Q. Since for every $\alpha \in B$ we have $\alpha + H_i \subseteq B$, the set B is open in T, and, therefore, ρ is a homeomorphism. ∎

Now, let $(\widehat{Q}_I, \widehat{T}_I)$ be the completion of (Q, T_I) and let $\widehat{A}_I$ be the closure of A_I in $\widehat{Q}_I$. It is well known that $\widehat{Q}_I$ has zero divisors, and in this case the triple $(\widehat{Q}_I, \widehat{T}_I, \widehat{A}_I)$ cannot be a representation of any tl-group. Analogously as we did for a topological field Q, we may consider the completion of a tl-group (G, F), where $G = Z^{(I)}$. It is well known that this completion $(\widehat{G}, \widehat{F})$ is a tl-group. The natural question arising here is whether $(\widehat{G}, \widehat{F})$ admits a representation. To tell the truth, we cannot solve this question as stated here. On the other hand, if we somewhat modify the definition of a representation we are able to answer affirmatively this question and, moreover, we may exhibit the relation between $\widehat{G}$ and $\widehat{Q}_I$. To do it, we say that a tl-group (G, F) admits a *general representation* (K, T, A) (in symbol : $(G, F) \sim$ $\sim (K, T, A)$), if K is a ring (with possible zero divisors), A is a subring of K such that K is the total quotient ring of A, T is a ring topology on K such that

$(U(K), T \mid U(K))$ is a topological group with $U(A)$ as a closed subgroup and the factor topological group $G(A) = U(K)/U(A)$ is a *tl*-group (with ordering defined by $(U(K)/U(A))_+ = A^r/U(A)$, where A^r is the set of regular elements of A, therefore, a *po*-group $G(A)$ is a *value group* of A in the sense of Mott, Schexnayder [95]) which is *tl*-isomorphic with G.

In a sequel we use a method of nonstandard analysis introduced by A. Robinson and, especially, we employ a variation of the nonstandard analysis introduced by E. Zakon [137], since it requires only the rudiments of firstorder logic.

Our principal goal is to prove the following theorem which fully describes the relation between $\widehat{Q}_I$ and $\widehat{G}$.

11.11. THEOREM. $(\widehat{G}, \widehat{F}) \sim (\widehat{Q}_I, \widehat{T}_I, \widehat{A}_I)$.

The proof of this theorem will be a consequence of several independent propositions which describe structures of $\widehat{Q}_I$ and $\widehat{G}$, respectively, and it will occupy the rest of this section. Included solely for the convenience of the reader, we introduce the basic facts about enlargement, which we frequently use.

For any set $A = A_0$ of individuals, the *superstructure* on A is the set $\boldsymbol{A} = \bigcup_{n \in N} A_n$, where N is the set of natural numbers and A_{n+1} is the set of all subsets of $A_0 \cup A_n$. The first order logic language $\boldsymbol{L}$ we need here is a simple modification of a clasical one, namely, we assume that all constants of $\boldsymbol{L}$ are in one-to-one correspondence with elements of $\boldsymbol{A}$ and identify the constants with the corresponding elements. Well-formed formulae (WFF) and sentences (WFS) are defined as usual with the restriction that all quantifiers must have form $(\forall x \in C)$ or $(\exists x \in C)$ with C a constant (i.e. $C \in \boldsymbol{A}$). Now, let A, B be two sets with superstructures $\boldsymbol{A}$, $\boldsymbol{B}$, respectively, and let

$$* : \boldsymbol{A} \longrightarrow \boldsymbol{B}$$

be a map of $\boldsymbol{A}$ into $\boldsymbol{B}$. We write $^*C$ for $^*(C)$. Let $^*\boldsymbol{A} = \bigcup_{n \in N} {}^*A_n$ (since $A_n \in \boldsymbol{A}$). Given a WFF α, we denote by $^*\alpha$ the formulae obtained from α by replacing in it each constant $C \in \boldsymbol{A}$ by $^*C$. Elements of the form $^*C$ $(C \in \boldsymbol{A})$ are called *standard*, their elements are called *internal*. Elements of $\boldsymbol{B}$ which are not internal are called *external*. A map $* : \boldsymbol{A} \longrightarrow \boldsymbol{B}$ which is one-to-one is called a *strict monomorphism* if

1) $*\phi = \phi$,
2) for every $y \in *A$, $y \subseteq *A$ holds ,
3) for every WFS α, $A \models \alpha$ iff $B \models *\alpha$.

A binary relation R in A is said to be concurrent if for any finite number of elements $a_1, \ldots, a_m \in D_1(R) = \{x : (\exists y)\,(x, y) \in R\}$, there exists b such that $(a_K, b) \in R$ for $K = 1, \ldots, m$. Then a strict monomorphism $* : A \longrightarrow B$ is called *enlarging* and $*A$ an *enlargement* of A if, for each concurrent relation R in A there is some $b \in *A$ such that $(*a, b) \in *R$ for all $a \in D_1(R)$ simultaneously.

If $*A$ is an enlargement of A, where A is the superstructure of A, we say frequently, that $*A$ $(\in *A)$ is an *enlargement* of A $(\in A)$. For any $X \subseteq A$ we may consider X as a subset of $*X$ and, furthermore, for any binary relation $R \subseteq X \times Y$, $X, Y \subseteq A$, we have $R \subseteq *R$.

Now, let K be the topological field with a topology $T = \sup(T_w : w \in \Omega)$ and let K be the superstructure on $K_0 = K \cup \Omega \cup \bigcup_{w \in \Omega} G_w$, $K = \bigcup_{n \in N} K_n$, and let $*K$ be an enlargement of K. Using property (3), it may be proved that $*K$ $(\in *K)$ is a field, $K \subset *K$ is a subfield and $*w$ $(w \in \Omega \in K)$ is a valuation of $*K$ with a value group $*G_w$ such that the diagram

$$\begin{array}{ccc} *K & \xrightarrow{\;*w\;} & G_w \cup \{\infty\} \\ \uparrow \cup & & \uparrow \cup \\ K & \xrightarrow{\;w\;} & G_w \cup \{\infty\} \end{array}$$

commutates. Let cG_w be the convex closure of G_w in $*G_w$ and let w be a valuation of $*K$ completing the diagram

$$\begin{array}{ccc} *K & \xrightarrow{\;*w\;} & *G_w \cup \{\infty\} \\ & {\scriptstyle w} \searrow \quad \swarrow {\scriptstyle \text{nat}} & \\ & *G_w / cG_w \cup \{\infty\} & \end{array}$$

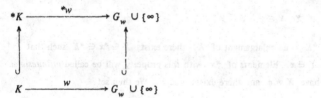

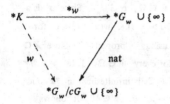

Let M_w be the maximal ideal of R_w and let

$$M = \bigcap_{w \in \Omega} M_w .$$

Then M is a subgroup of $({}^*K, +)$ and on the factor group ${}^*K/M$ we may define a topology such that a subbase B of the neighbourhoods of 0 consists of the sets

$$U_{{}^*w,\alpha}/M = {}^*(U_{w,\alpha})/M = \{x + M \in {}^*K/M : {}^*w(x) > \alpha\},$$

where $w \in \Omega$, $\alpha \in G_w^+$. Clearly, ${}^*K/M$ is then a topological group.

Now, under the injection $x \longmapsto x + M$, $x \in K$, we may identify K with a subgroup in ${}^*K/M$. Let $\overline{K}$ be the closure of K in ${}^*K/M$. Then the following theorem holds.

11.12. **THEOREM.** $\overline{K}$ *is a topological ring homeomorphic with the completion* $\widehat{K}$ *of* (K, T).

P r o o f. At first, we may consider $\widehat{K}$ to be a factor set of a set of all Cauchy filters $\boldsymbol{x}$ under the relation $\boldsymbol{x} \sim \boldsymbol{y}$ iff $\boldsymbol{x} \cap \boldsymbol{y}$ is a Cauchy filter. Let $\overline{x} \in \widehat{K}$ be such that $\boldsymbol{x} \in \overline{x}$.

Now, let $\overline{x} \in \widehat{K}$. Since the binary relation R such that

$$(X, Y) \in R \text{ iff } X, Y \in \boldsymbol{x},\ X \subseteq Y,$$

is concurrent and ${}^*K$ is an enlargement of K, there exists $X \in {}^*\boldsymbol{x} \in {}^*K$ such that $X \subseteq {}^*Y$ for each $Y \in \boldsymbol{x}$. Elements of ${}^*\boldsymbol{x}$ with this property will be called *infinitesimal.* Since $\phi \notin \boldsymbol{x}$, we have $X \neq \phi$ and there exists $y \in X$. We then set

$$\rho : \widehat{K} \longrightarrow \overline{K} \subseteq {}^*K/M ,$$
$$\rho(\overline{x}) = y + M .$$

This definition is correct. In fact, let $z \in X$ and let $w \in \Omega$, $\alpha \in G_w$. Since $\boldsymbol{x}$ is a Cauchy filter, there exists $Y \in \boldsymbol{x}$ such that $w(Y - Y) > \alpha$ (i.e. $w(u - t) > \alpha$ for each $u, t \in Y$) and it follows ${}^*w({}^*Y - {}^*Y) > \alpha$, using property 3). Since $X \subset {}^*Y$, we have ${}^*w(y - z) > \alpha$, and $y - z \in M_w$ for every $w \in \Omega$. Thus, $y + M = z + M$. Now, let $Z \in {}^*\boldsymbol{x}$ be infinitesimal, then $X \cap Z$ is infinitesimal in ${}^*\boldsymbol{x}$ and for any $z \in Z$ there exists $u \in X \cap Z$ such that $y - u,\ u - z \in M$. It follows that $y + M = z + M$. Finally, it is easy to see that if $z \in \bar{x}$, Z is infinitesimal in $*z$ and $z \in Z$, then $y + M = z + M$.

Further, we show that $\rho(\bar{x}) \in \bar{K}$. Clearly, $\{\{t + M : t \in Y\} : Y \in x\}$ is a base in K of a filter F in $*K/M$. Let $w_1, \dots, w_m \in \Omega$, $\alpha_i \in G^+_{w_i}$. Then we may find $Y \in x$ such that

$$*w_i(*Y - *Y) > \alpha_i, \quad i = 1, \dots, m.$$

Thus, $*w_i(y - Y) > \alpha_i$ for every i, and, $(y + \bigcap_{i=1}^{m} *U_{w_i, \alpha_i})/M \in F$. Therefore, $y + M = \lim F$ and $y + M \in \bar{K}$.

The map ρ is injective. Indeed, let $\rho(\bar{x}) = x + M = y + M = \rho(\bar{y})$, where $x(y)$ is an element of an infinitesimal element $X(Y)$ of $*x$ $(*y)$. Let $w \in \Omega$, $\alpha \in G_w$. Then there exist $A \in x$, $B \in y$ such that $A - A \subseteq U_{w,\alpha}$, $B - B \subseteq U_{w,\alpha}$. Then $A \cup B \in x \cap y$ and for any $a \in A$, $b \in B$ we have

$$w(a - b) \geqq \min(*w(a - x), *w(x - y), *w(y - b)) > \alpha.$$

Thus, $\bar{x} = \bar{y}$.

Further, let $y + M \in \bar{K}$. Then there exists a filter F in $*K/M$ with a base B in K such that $y + M = \lim F$. Let x be a filter in K with a base B, then it is easy to see that x is a Cauchy filter and $\rho(\bar{x}) = y + M$.

It is easy to see that ρ is a group isomorphism, where for $\bar{x}, \bar{y} \in \hat{K}$, $\bar{x} + \bar{y} = \bar{z}$ iff z is a (Cauchy) filter with a base $x + y = \{X + Y : X \in x, Y \in y\}$.

Now, let $R = \{x \in *K : x + M \in \bar{K}\}$. Then R is a subring in $*K$, and, since $R \subseteq R_w$ for every $w \in \Omega$, it follows that M is an ideal of R. Thus, we may identify the group $(\bar{K}, +)$ with the factor group $(R/M, +)$ and, moreover, on $\bar{K}$ we may define a multiplication as in the factor ring R/M and, clearly, $\bar{K}$ is a ring isomorphic with the ring K, where $\bar{x} \,.\, \bar{y} = \bar{z}$ iff z is a filter with a base $x \,.\, y = \{X \,.\, Y : X \in x, Y \in y\}$.

Finally, we show that ρ is a homeomorphism. The base of the neighbourhoods of zero in $\bar{K}$ consists of the sets

$$[\bigcap_{i=1}^{m} U_{w_i}, \alpha] = \{\bar{x} \in \hat{K} : \bigcap_{i=1}^{m} U_{w_i, \alpha_i} \in x\},$$

where $m \in N$, $w_i \in \Omega$, $\alpha_i \in G^+_{w_i}$. We consider $\bar{K}$ to be a topological ring with a topology induced from $*K/M$. Then we show that

$$\rho([\bigcap_{i=1}^{m} U_{w_i,\alpha_i}]) = \bigcap_{i=1}^{m} *U_{w_i,\alpha_i}/M \cap \bar{K}.$$

In fact, let $\bar{x} \in \hat{K}$ be such that $\cap U_{w_i,\alpha_i} \in x$, $\rho(\bar{x}) = y + M$, where $y \in X \in *x$ for an infinitesimal X. Let φ be a WFS such that

$$\varphi \equiv (\forall a \in K)(a \in U_{w_1,\alpha_1} \wedge \ldots \wedge a \in U_{w_m,\alpha_m} \Leftrightarrow a \in \bigcap_{i=1}^{m} U_{w_i,\alpha_i}.$$

Since $K \models \varphi$, we have $*K \models *\varphi$ and it follows

$$y \in X \subseteq *(\bigcap_{i=1}^{m} U_{w_i,\alpha_i}) = \bigcap_{i=1}^{m} *U_{w_i,\alpha_i}.$$

Conversely, let $y + M \in \cap *U_{w_i,\alpha_i}/M \cap \bar{K}$, $\rho(\bar{x}) = y + M$, where $y \in X \in *x$ for infinitesimal X; hence $X \subseteq \cap *U_{w_i,\alpha_i}$. Since

$$K \models (\forall a \in x)(\forall y \in K_1)(a \subseteq y \Rightarrow y \in x),$$

using property (3) we obtain that $\cap *U_{w_i,\alpha_i} \in *x$. Hence, if we denote φ_1 a WFS such that

$$\varphi_1 \equiv \bigcap_{i=1}^{m} U_{w_i,\alpha_i} \in x,$$

we have $*K \models *\varphi_1$ and it follows $K \models \varphi_1$. Therefore, $\bar{x} \in [\cap U_{w_i,\alpha_i}]$ and, clearly, ρ is a homeomorphism. ∎

For the further investigation we need to describe a structure of the completion $\hat{R}_w$ for $w \in \Omega$ in the language of nonstandard analysis.

11.13. **PROPOSITION.** *The completion of* R_w *is* $*R_w/M \cap \bar{K}$ *for every* $w \in \Omega$.

Proof. We show that $*R_w/M \cap \bar{K}$ is the closure of R_w in $\bar{K}$. In fact, let $x + M \in \bar{R}_w$ (the closure of R_w) and let us assume that $x + M \notin *R_w/M$. Since $M \subseteq *R_w$, we have $x \notin *R_w = R_{*w}$ (using 3)) and it follows $*w(x) < 0$. Moreover, there exists z such that

$$z \in R_w \cap ((x + M) + *U_{w,0}/M;$$

for such a z we have $*w(z - x) > 0$, $*w(z) = w(z) \geq 0$, a contradiction.

Conversely, let $x + M \in {}^*R_w/M \cap \bar{K}$. Then ${}^*w(x) \geq 0$ and if we suppose that $x + M \notin \bar{R}_w$, there exist $w_1, \ldots, w_m \in \Omega$ such that

$$((x + M) + \bigcap_{i=1}^{m} {}^*U_{w_i,\alpha_i}/M) \cap R_w = \phi .$$

We may assume without lost of generality that $w = w_1$. Since $x + M \in \bar{K}$, there exists $a \in K$ with the property

$$ {}^*w_i(x - a) > \alpha_i, \quad i = 1, \ldots, m .$$

Then ${}^*w_i(a) < 0$ and ${}^*w(x - a) = {}^*w_1(x - a) = {}^*w_1(a) < 0$, a contradiction. ∎

We are now able to describe a (Manis) valuations $\tilde{w}$ which are the continuous extension of w on a valuation of a ring $\hat{K}$.

11.14. PROPOSITION. *Let* $w \in \Omega$ *and let* $\tilde{w} : \hat{K} \longrightarrow G_w \cup \{\infty\}$ *be such that*

$$\tilde{w}(x + M) = \begin{cases} \infty & \text{iff } {}^*w(x) \in {}^*G_w^+ - G_w, \\ {}^*w(x), & \text{otherwise} . \end{cases}$$

Then $\tilde{w}$ *is the continuous extension of* w *and* $R_{\tilde{w}}$ *is the completion of* R_w.

Proof. Clearly, $\tilde{w}(x + M) \in G_w \cup \{\infty\}$ and $\tilde{w}$ is a (Manis) valuation of K such that $R_{\tilde{w}} = {}^*R_w/M \cap \hat{K} = \bar{R}_w$ by 11.13. ∎

11.15. COROLLARY. *The topology on* K *equels* $\sup(T_{\tilde{w}} : w \in \Omega)$.

Proof. The base of the neighbourhoods of zero in $\hat{K}$ $(= \bar{K})$ consist of the sets $\cap {}^*U_{w_i,\alpha_i}/M \cap \bar{K} = \cap U_{\tilde{w}_i,\alpha_i}$ by 11.12 and 11.14. ∎

Now, let $K = \bigcup_m K_m$ be the superstructure on the field $Q = K_0$. Let $P_I(\Omega_I)$ be the set of all *i*th prime numbers p_i (p_i-adic valuations of Q) for $i \in I$ and let φ be a WFS such that

$$\varphi \equiv (\forall p \in N)(\forall x \in N)(\forall y \in N)(p \in P_I \Leftrightarrow (p \neq 1 \wedge (p = x \cdot y \Rightarrow x = 1 \vee y = 1)) .$$

Then $K \models \varphi$ states that P_I is a set of prime numbers in N. Since ${}^*K \models {}^*\varphi$, the set $*P_I$ is a set of 'prime numbers' in $*N$. Analogously, let μ be a WFS which states that for every $p \in P_I$ there exists a p-adic valuation w_p of Q whith a value group. Z. Since $K \models \mu$, we have $*K \models *\mu$ and it follows that for every $p \in *P_I$ there exists a 'p-adic' valuation w_p of field $*Q$ with a value group $*Z$. Clearly, for $p \in P_I \subset *P_I$ we have $w_p = *w_p$.

For our purpose we need the following proposition, the proof of which may be found in Močkoř [88].

11.16. **PROPOSITION.** $\widehat{A}_I = \bigcap_{i \in I} R_{\tilde{w}_i}$.

The following proposition describes fully the set $U(\widehat{Q}_I)$. The elements of $\widehat{O}_I$ we denote by x $(= x + M)$, where $x \in *Q$.

11.17. **PROPOSITION.** *Let* $x \in *Q$. *Then* $x \in U(\widehat{Q}_I)$ *if and only if* $*w_i(x) \in$ $\in Z$ *for each* $i \in I$.

Proof. Let $x \in U(\widehat{Q}_I)$. If there exists $i \in I$ such that $*w_i(x) = \omega$ for some $\omega \in *N - N$, then for $y \in \widehat{Q}_I$ such that $x \cdot y = 1$ we have $*w_i(x \cdot y - 1) \in *N - N$ and it follows $*w_i(y) = -\omega$. Then for any $z \in Q$ we have

$$-\omega = *w_i(y - z) < n, \quad n \in N,$$

and $y \notin \widehat{Q}_I$, a contradiction.

Conversely, without lost of generality we may suppose that $*w_i(x) = -a_i \leqq 0$ for each $i \in I$. Then by 11.14, $\tilde{w}_i(z) = *w_i(z) \geqq 0$, $i \in I$, where $z = x^{-1}$ and by 11.16, $z \in \widehat{A}_I \subseteq \widehat{Q}_I$. Hence for every pair $(i, a) \in I \times N$ there exists $y_{i,a} \in A_I$ such that

$$*w_i(z - y_{i,a}) > a + 2a_i.$$

Since $*w_i(z) = a_i < a + 2a_i$, we have $y_{i,a} \neq 0$ and $y_{i,a}^{-1} \in Q$, $a_i = *w_i(z) = w_i(y_{i,a})$. Then we obtain

$$*w_i(x - y_{i,a}^{-1}) = *w_i(x(y_{i,a} - z)y_{i,a}^{-1}) > -a_i + a + 2a_i - a_i = a.$$

Therefore, we have proved

$$(\forall (i, a) \in I \times N)(\exists z_{i,a} \in Q)(*w_i(x - z_{i,a}) > a).$$

Now, let $i_1, \ldots, i_n \in I$, $a_1, \ldots, a_n \in N$. Using the approximation theorem for valuations of Q we may find an element $y \in Q$ such that

$$w_{i_t}(y - z_{i_t, a_t}) > a_t, \quad t = 1, \ldots, n.$$

Hence,

$${}^*w_{i_t}(x - z) = {}^*w_{i_t}(x - z_{i_t, a_t} + z_{i_t, a_t} - y) > a_t, \quad t = 1, \ldots, n,$$

and it follows $x \in \widehat{Q}_I$. Clearly, $x \cdot y = 1$ in $\widehat{Q}_I$ and $x \in U(\widehat{Q}_I)$. ■

To show that $(\widehat{Q}_I, \widehat{T}_I, \widehat{A}_I)$ is a general representation we have to prove that $U(\widehat{Q}_I)$ with an induced topology is a topological group (and not only a topological semigroup).

11.18. PROPOSITION. *$(U(\widehat{Q}_I), \cdot, \widehat{T}_I \mid U(\widehat{Q}_I))$ is a topological group.*

Proof. We show that a map $x \longmapsto x^{-1}$ is continuous in $U(\widehat{Q}_I)$. In fact, let $U = (1 + \bigcap_{t=1}^{m} U_{\widetilde{w}_{i_t}, a_t}) \cap U(\widehat{Q}_I)$ be an arbitrary neighbourhood of 1. Since (Q, T_I) is a topological field, there exists a neighbourhood $V = (1 + \bigcap_{s=1}^{n} U_{w_{j_s}, b_s}) \cap Q^*$ of 1 in Q^* such that

$$V^{-1} \subseteq 1 + \bigcap_{t=1}^{m} U_{w_{i_t}, a_t} = U.$$

Let $z \in V = (1 + \bigcap_s U_{\widetilde{w}_{j_s}, b_s}) \cap U(\widehat{Q}_I)$, then by 11.17, ${}^*w_i(z) \in Z$ for every $i \in I$ and ${}^*w_{j_s}(z - 1) > b_s$, $s = 1, \ldots, n$. Without lost of generality we may assume that $\{w_{j_1}, \ldots, w_{j_n}\} \cap \{w_{i_1}, \ldots, w_{i_m}\} = \phi$. Since $z \in \widehat{Q}_I$, there exists $x \in Q$ such that

$$\begin{aligned} &{}^*w_{j_s}(z - x) > b_s, \quad s = 1, \ldots, n, \\ &{}^*w_{i_t}(z - x) > \max(a_t + 2{}^*w_{i_t}(z), {}^*w_{i_t}(z)) \in Z, \quad t = 1, \ldots, m. \end{aligned}$$

Since $w_{j_s}(x - 1) = {}^*w_{j_s}(x - z + z - 1) > b_s$, we have $x \in V$ and $x^{-1} \in U$. Then we obtain

$${}^*w_{i_t}(z^{-1} - 1) = {}^*w_{i_t}(z^{-1} - x^{-1} + x^{-1} - 1) > a_t, \quad t = 1, \ldots, m,$$

and $V^{-1} \subseteq U$. ■

Now, the same method of enlargement we may use for investigation of properties of the completion $\widehat{G}$ of $(Z^{(I)}, F)$. As in a case of topological fields, we may do it in a more general way.

So, let G be a *tl*-group with a subbase $\boldsymbol{H}$ of zero consisting of prime ideals, $\boldsymbol{H} = \{H_i : i \in I\}$. Let $\hat{G}$ be the superstructure on the set $G_0 = G \cup I$ and let $^*G$ be an enlargement of G. Let

$$H = \bigcap_{i \in I} {}^*H_i \ .$$

Then H is an *o*-ideal of $^*G$ and in a group $^*G$ we may define a topology in such a way that $\{ {}^*H_i : i \in I\}$ is a subbase of the neighbourhood of zero. Clearly, $^*G$ is a *tl*-group and since H is a closed *l*-ideal of $^*G$, we may consider a factor *tl*-group $^*G/H$. Then the canonical map

$$G \longrightarrow {}^*G/H$$

is an injection as follows from the fact $^*H_i \cap G = H_i$, $i \in I$. Then the following proposition is an analogue of 11.12.

11.19. **PROPOSITION.** *The closure cG of G in $^*G/H$ is tl-isomorphic with the completion $\widehat{G}$ of G.*

P r o o f. At first, $\widehat{G}$ may be considered to be the factor set of the set of all Cauchy filters in G. Elements of this factor set will be denoted by α, their elements (i.e. Cauchy filters) by $\boldsymbol{\alpha}, \boldsymbol{\beta}$ etc. Then $\boldsymbol{\alpha}, \boldsymbol{\beta} \in \overline{\gamma}$ iff $\boldsymbol{\alpha} \cap \boldsymbol{\beta}$ is a Cauchy filter. The base of the neighbourhoods of zero in $\widehat{G}$ consists of the sets

$$[\bigcap_{i=1}^{m} H_i] = \{\overline{\alpha} : \bigcap_{i=1}^{m} H_i \in \boldsymbol{\alpha}\}, \ H_i \in \boldsymbol{H} \ .$$

The operations in $\widehat{G}$ are defined analogously as in the proof of 11.12, and analogously we may find for every $\alpha \in \widehat{G}$ an infinitesimal element $X \in {}^*\boldsymbol{\alpha}$, $X \neq \phi$. Then we define a map ρ as follows :

$$\rho : \widehat{G} \longrightarrow cG \ ,$$

$$\rho(\overline{\alpha}) = \alpha + H \ , \ \text{ where } \ \alpha \in X \ .$$

Using the proof of 11.12, it may be done similarly so that this definition is correct and that ρ is a *tl*-isomorphism. ∎

The following proposition is an analogy of 11.15.

11.20. **PROPOSITION.** *For every* $i \in I$, $\widehat{H}_i = {}^*H_i/H \cap G$ *is the closure of* H_i *in* G. *The set* $\{\widehat{H}_i : i \in I\}$ *is a realizator of* $\widehat{G}$ *and it is a subbase of the neighbourhoods of zero in* G.

P r o o f. The first part of the proposition follows immediately from the fact that H_i is dense in ${}^*H_i/H \cap G$. It may be easily seen that $\widehat{H}_i$ is a prime *l*-ideal of $\widehat{G}$ and

$$\bigcap_{i \in I} \widehat{H}_i = \bigcap_{i \in I} ({}^*H_i/H \cap \widehat{G}) = \bigcap_{i \in I} ({}^*H_i/H) \cap \widehat{G} = \{0\} .$$

Moreover, the topology in $\widehat{G}$ is induced from the one in ${}^*G/H$, i.e. the subbase of the neighbourhoods of zero consists of the set $\{{}^*H_i/H \cap \widehat{G} : i \in I\}$. ∎

Now, we are able to prove 11.11. At first, using the nonstandard construction of $\widehat{G}$ we may fully describe elements of $\widehat{G}$. So, let $G = Z^{(I)}$ for $I \subseteq N$, card $I = \aleph_0$, and let ${}^*G$ be an enlargement of G, $\alpha \in {}^*G$. Then $\alpha + H \in \widehat{G}$ if and only if $\alpha_i \in Z$ for every $i \in I$. This follows immediately from 11.20, where ${}^*H_i = \{\beta \in {}^*G : \beta_i = 0\}$, $i \in I$. Let $\widehat{w}$ be a semivaluation associated with a ring $\widehat{A}_I$, i.e.

$$\widehat{w} : U(\widehat{Q}_I) \longrightarrow U(\widehat{Q}_I)/U(\widehat{A}_I) ,$$

and let $x \in U(\widehat{Q}_I)$. According to 11.17, ${}^*w_i(x) \in Z$ for every $i \in I$. Moreover, interpreting in ${}^*Q$ a suitable WFS, we may find an element $i_0 \in {}^*I$ such that $w_i(x) = 0$ for every $i \in {}^*I$, $i > i_0$. Since

$${}^*G = \{\alpha \in {}^*Z^{*I} : \alpha \text{ is internal and there exists } i_0 \in {}^*I \text{ such that } \alpha_i = 0, \text{ for every } i \in {}^*I,\ i > i_0\},$$

we may find an element $\alpha \in {}^*G$ such that

$\alpha_i = {}^*w_i(x) \in \mathbf{Z},\ i \in I.$

We define a map ρ as follows:

$$\rho : U(\hat{Q}_I)/U(\hat{A}_I) \longrightarrow \hat{G}$$
$$\rho(w(x)) = \alpha + H.$$

If x and y are elements of ${}^*Q$ such that ${}^*w_i(x) = {}^*w_i(y) \in \mathbf{Z}$ for every $i \in I$, then we have $\tilde{w}_i(x) = \tilde{w}_i(y)$ according to 11.14, and since $x, y \in U(\hat{Q}_I)$, we have $x \cdot y^{-1},\ x^{-1} \cdot y \in \bigcap_{i \in I} R_{\tilde{w}_i} = \hat{A}_I$ by 11.16. It follows $w(x) = w(y)$ and the definition is correct. It is clear that ρ is an o-isomorphism. Since $\rho^{-1}(\hat{H}_i) = U(R_{\tilde{w}_i}),\ i \in I$, ρ is open and continuous, hence, ρ is a homeomorphism. Since $U(\hat{Q}_I)/U(\hat{A}_I)$ is a topological group, it is a tl-group which is tl-isomorphic with $\hat{G}$ and the theorem is proved.

12. APPLICATIONS OF GROUPS OF DIVISIBILITY

The idea of basic applications of groups of divisibility consists of the following method: To formulate the ring-theoretic problem in terms of group of divisibility (if it is possible), solve the problem there and pull back the solution whenever possible to the integral domains. The main tool in the pull-back process is theorem 8.1. However this theorem reduces the class of admissible domains to the class of GCD-domains.

The method considered before was first used by Nakayama [104], who showed that a completely integrally closed domain need not be an intersection of rank 1 valuation rings, which disproves the conjecture of Krull [79] and Clifford [23]. His proof is rather complicated, although a more simple version was carried out by J. Ohm (see R. Gilmer [42]). Ohm showed that there is an infinite dimensional Bezout domain that is completely integrally closed and does not admit a rank 1 valuation. This leads naturally to the question of whether this is possible for a domain with a finite dimension. The answer is positive as the following example shows with the aid of groups of divisibility.

12.1. **EXAMPLE** (Sheldon [117]). There exists a Bezout domain A such that $\dim A = 2$, A is completely integrally closed and A is not an intersection of rank 1 valuation rings.

CONSTRUCTION. Let G_0 be the set of all functions f from $[0, 1]$ into $\mathbf{Z}$ where we identify two functions if they agree on all but finitely many points. Let F be the set of all step functions of G_0, i.e.

$$F = \{f \in G_0 : \text{there exist } 0 = a_0 < a_1 < \ldots < a_n = 1 \text{ such that for all but finitely many } x, y \in (a_i, a_{i+1}) \text{ one has } f(x) = f(y)\}.$$

For $b \in [0, 1]$ we define $h_b \in G_0$ such that

$h_b(b) = 1$,
$h_b(x) = [1/(x-b)^2]$, $x \in [0, 1] - \{b\}$,

where $[\cdots]$ denotes the greatest integer function. Let G be a subgroup of G_0 generated by $F \cup \{h_b\}_{b \in [0,1]}$. Thus, a typical element of G is of the form

$$(12.2) \qquad f + n_1 h_{b_1} + \ldots + n_r h_{b_r}, \quad f \in F,\ n_i \in \mathbf{Z},\ b_i \in [0, 1].$$

Then G may be ordered by

$$g \geqq g' \text{ iff } g(x) \geqq g'(x) \text{ for all but a finitely many } x \in [0, 1].$$

Clearly, G is an l-group, and by 8.1,. there exists a Bezout domain A such that $G(A) \cong_0 G$. We set

$P_b = \{g \in G_+ : h_b$ appears with a non zero coefficient in the expression (12.2) for $g\}$,
$P'_b = \{g \in G_+$ there exists $c \in [0, b)$ such that $g(x) > 0$ for all but finitely many $x \in (c, b)\}$,
$P''_b = \{g \in G_+ :$ there exists $c \in (b, 1]$ such that $g(x) > 0$ for all but finitely many $x \in (b, c)\}$.

Then $\{\phi\} \cup \{P_b : b \in [0, 1]\} \cup \{P'_b : b \in (0, 1]\} \cup \{P''_b : b \in [0, 1)\}$ is the set of prime filters of G_+ and the only containment relation among them are $\phi \subset P_b \subset P'_b$, $P_b \subset P''_b$. By YAKABE's bijection, $\dim A = 2$.

Now, let K be the quotient field of A and let $f_0 \in F \cap G_+$ and let us choose $x_0 \in K$ such that $w_A(x_0) = -f_0$ (i.e. $x_0 \notin A$). If R is a valuation domain of K, $\dim R = 1$, $A \subsetneqq R$, then $w_A(M_R \cap A^*) = P_b$ for some $b \in [0, 1]$. Thus, $x_0 \in R$ and $x_0 \in \cap \{R : R$ is a rank 1 valuation domain of K, $A \subsetneqq R\} - A$, and so A is not an intersection of rank 1 valuation domains.

Another example involving complete integral closure deals with the idempotency of the complete integral closure. More exactly, by $c(A)$ we denote the complete integral closure of A in the quotient field. Then $c^{n+1}(A) = c(c^n(A))$. Gilmer and Heinzer [46] proved that there is a domain A such that $c^2(A) \neq c(A)$. The following example of P. Hill [61] which uses the group of divisibility shows much more.

12.3. EXAMPLE. There exists a Bezout domain A such that $c^n(A) \subsetneqq c^{n+1}(A)$ for all $n \in N$.

CONSTRUCTION. Let $H = Z \oplus_L Z$. View an element of H^{Z^*} as $(\ldots (a_{-1}, b_{-1}), (a_1, b_1), \ldots) = ((a_n, b_n)), a_n, b_n \in Z, n \in Z^* = Z - \{0\}$. Let us define

$$G_1 = \{((a_n, b_n)) \in H^Z : a_n = 0 \text{ for } n < 0,\ a_n = 0 \text{ for all but finitely many } n \in Z, \text{ and } b_n = 0 \text{ for all but finitely many } n < 0\}.$$

Giving H^{Z^*} the product ordering and considering G_1 as a subgroup, G_1 becomes an l-group. Let $\{N_k\}_{k \in N}$ be a countably infinite partition of N such that N_k is infinite for all but finitely many $k \in N$. We define

$$G_2 = \{((a_n, b_n)) \in G_1 : b_n \text{ is constant on } N_k \text{ for all but finitely many } k \in N, \text{ and for all } k \in N \text{ there exist } a, b \in Z \text{ such that } b_n = na + b \text{ for all but finitely many } n \in N_k\},$$

$$K = \{((a_n, b_n)) \in G_1 : a_n = 0 \text{ for all } n \in Z, b_n = 0 \text{ for all } n < 0 \text{ and } b_n = 0 \text{ for all but finitely many } n \in Z\}.$$

Then $K \subset G_2$ are subgroups of G_1 and $G_2/K \cong_o G_1$. Let

$$\beta : G_2 \longrightarrow G_1$$

be the composition of the natural o-homomorphism and the o-isomorphism. For $n \in N$, $n \geqq 2$, we define

$$G_{n+1} = \beta^{-1}(G_n).$$

Then $G_{n+1}/K \cong_o G_n$.

To finish this construction we need a certain consideration.

J. L. Mott [92] has derived a group of divisibility of a complete integral closure of a Bezout domain in the following way:

For a *po*-group G, an element $g \in G$, $g \neq 0$, is said to be *bounded* if there is an element $h \in G$ such that $ng \leqq h$ for all $n \in N$. The set of all bounded elements forms a convex semigroup of G and generates an o-ideal $B(G)$

of G. Using the method of Sheldon [117], it is possible to show that if a complete integral closure $c(A)$ of A is a quotient ring of A, then $G(c(A)) \cong$ $\cong_0 G(A)/B(G(A))$. Since every overring of a Bezout domain A is a quotient ring of A, we have the following proposition.

12.3.1. PROPOSITION. *If A is a Bezout domain, G its group of divisibility, then $G(c(A)) \cong_0 G/B(G)$.*

We may return to the construction of 12.3. It is possible to show that for each positive integer n, $B(G_{n+1}) = K$. Thus, $G_{n+1}/B(G_{n+1}) \cong_0 G_n$, and

$$B(G_1/B(G_1)) = B(G_1/K) \neq \{0\}, \quad B((G_1/K)/B(G_1/K)) = \{0\} .$$

Let G be the o-sum of groups G_n, $n \in N$. By 8.1, there exists a Bezout domain A such that $G(A) \cong_0 G$ and it follows $c^n(A) \neq c^{n+1}(A)$.

The following example involving complete integral closure, has something to do with the power series rings. It is known (see Sheldon [120]) that if R is a valuation domain, then the quotient field $K(R[[X]])$ of $R[[X]]$ equals the quotient field $K(c(R)[[X]])$ of $c(R)[[X]]$. The example of Sheldon [120] shows that this cannot be extended from valuation domains to Bezout domains. To show Sheldon's construction, we first need the following two results.

If S is a multiplicative system of a domain A, then $K(A_S[[X]]) =$ $= K(A[[X]])$ if and only if for every sequence $s_1, s_2, s_3, \ldots$ of elements of S, $\bigcap_{n \in N} s_n A \neq \{0\}$ holds. Then $K(A[[X]] = K(c(A)[[X]])$ for a Bezout domain A if and only if every countable sequence of bounded elements of $G(A)_+$ has an upper bound as follows from the preceding result and from 12.3.1.

12.4. EXAMPLE. Let $T = Z \oplus_l Z$ and let G be the set of all functions from N into T with a finite support. Then G is an l-group with respect to the pointwise addition and ordering. The bounded positive elements are those satisfying $f(n) = (z_n, 0)$ for each n. The set of functions $\{t_1, t_2, \ldots\}$ defined by

$$t_i(n) = (0, 0), \quad i \neq n,$$
$$t_i(i) = (1, 0),$$

is a countable set of bounded positive elements of G and clearly, this set can have no upper bound in G. If A is a Bezout domain such that $G(A) = G$, $K(A[[X]]) \neq$ $\neq K(c(A)[[X]])$.

The further example involving group of divisibility concerns a *locally Noetherian domain*, i.e. a domain A for which all its localizations at maximal ideals are Noetherian. The obvious conjecture that a locally Noetherian domain is Noetherian is false. The first such counter example was given by Nakano without using the group of divisibility. The following example which uses this group and is more simpler that the original one,is due to Heinzer and Ohm [57].

12.5. EXAMPLE. There exists a locally Noetherian domain which is not Noetherian.

CONSTRUCTION. Let

$G = \{ f \in Z^N$: there exists $k \in Z$ such that $f(n) = k$ for all but a finitely many $n \in N\}$.

Then G is an l-group with component wise addition and ordering. Let A be a Bezout domain such that $G(A) = G$. For $i \in N$ we define

$M_i = \{ a \in A$: if $w_A(a) = f$ then $f(i) > 0\} \cup \{0\}$,

$M_\infty = \{a \in A$: if $w_A(a) = f$ then $f(n) > 0$ for all but finitely many $n \in N\}$.

Then $\{M_i\}_{i \in N \cup \{\infty\}}$ is the set of maximal ideals of A. Since $G(A_{M_i}) = Z$, it follows that A is locally Noetherian. M_∞ is an ideal of A which is not finitely generated, thus A is not Noetherian.

One of the most interesting examples of using Theorem 8.1 is the following one which is due to Lewis [75] and which concerns the spectrum of rings.

For a domain A we denote by Spec A the set of prime ideals of A considered to be a partially ordered set under the set inclusion.

If $(X, \leqslant)$ is a partially ordered set, the following two conditions are of interest:

(K1) *Every chain of* X *has a supremum and infimum,*

(K2) *If* $x, y \in X$, $x < y$, *then there exist* $x_1, y_1 \in X$ *such that* $x \leqslant x_1 < < y_1 \leqslant y$ *and there no element of* X *properly between* x_1 *and* y_1 *exists. (This is denoted by* $x_1 << y_1$.*)*

It is well known that Spec A satisfies (K1) and (K2).

A partially ordered set $(X, \leqslant)$ is called a *tree* if $x, y, z \in X$, $x \leqslant z$, $y \leqslant z$ imply $x \leqslant y$ or $y \leqslant x$. If A is a Bezout domain, then Spec A is a tree with a unique minimal element.

12.6. EXAMPLE. If $(X, \leqslant)$ is a tree satisfying (K1) and (K2) and X has a unique minimal element, then there exists a Bezout domain A such that Spec A is order isomorphic to X ($X \cong$ Spec A).

We show the original proof of this theorem.

Let us define

$Y = \{y \in X :$ there exists $z \in X$ such that $z << y\}$ and let

$G = \{f \in \mathbf{Z}^Y : f(y) = 0$ for all but finitely many $y \in Y\} = \mathbf{Z}^{(Y)}$.

With pointwise addition G is a group. For $f \in G$ let

$$M(f) = \{y \in Y : f(y) \neq 0 \text{ and } f(s) = 0 \text{ for all } s \in Y, s < y\}.$$

We define

$$G_+ = \{f \in G : f(y) > 0 \text{ for all } y \in M(f)\}.$$

It is possible to show without any difficulty that G is a *po*-group with a positive cone G_+. Moreover, it should be observed that for $f, g \in G$,

$$f \geqslant g \Leftrightarrow \forall y \in Y\ ((y < x \Rightarrow f(y) = g(y)) \Rightarrow f(x) \geqslant g(x)) \text{ for all } x \in Y.$$

An element $f \in G_+$ such that $M(f)$ consisting of exactly one element is called irreducible. At first it is necessary to prove that G is an *l*-group.

12.6.1. LEMMA. *G is an l-group and for every* $f \in G_+$, $f \neq 0$, *there exists a unique set* $\{f_1, \ldots, f_n\}$ *of pairwise disjoint irreducible elements of* G *such that* $f = f_1 + \cdots + f_n$.

P r o o f. Let $f, g \in G$. Then since Y is a tree ($X \supseteq Y$ is a tree), for every $x \in Y$ there is at most one $y \in M(f-g)$ such that $y \leqslant x$. Then we define an element $h \in G$ as follows.

(i) $h(x) = f(x) = g(x)$ if there does not exist an $y \in M(f-g)$ such that $y \leqslant x$;

(ii) $h(x) = f(x)$ if there exists $y \in M(f-g)$, $y \leqslant x$, $f(y) < g(y)$;

(iii) $h(x) = g(x)$ if there exists $y \in M(f-g)$, $y \leqslant x$, $g(y) < f(y)$.

Then it is easy to see that $h = f \wedge g$ in G and G is an l-group.

Now, let $f \in G_+$, $f \neq 0$, and let $M(f) = \{x_1, \ldots, x_n\}$. Then for $i = 1, \ldots, n$ we define

$$f_i(x) = \begin{cases} f(x), & x \geqslant x_i,\ x \in Y; \\ 0, & x \ngeqslant x_i,\ x \in Y. \end{cases}$$

It is possible to show that the set $\{f_1, \ldots, f_n\}$ satisfies the conditions of the lemma. ∎

Now, let A be a Bezout domain such that $G(A) \cong_0 G$. According to 2.5, there is a bijection between the set of prime ideals of A (i.e. Spec A) and the set $F(G)$ of all prime filters in G_+ and this bijection preserves ordering by inclusion. Hence, it suffices to construct an order-preserving bijection between $(F(G), \subseteq)$ and $(X, \leqslant)$. For $x \in X$ we define

$$F_x = \{f \in G_+ : \text{there exists } y \in M(f) \text{ such that } y \leqslant x\}.$$

It is easy to see that F_x is a prime filter in G_+, i.e. $F_x \in F(G)$. Hence, we obtain a map

$$F : X \longrightarrow F(G)$$

such that $F(x) = F_x$. It is immediately seen from the definition that F is order-preserving. Let $x, x' \in X$, $x, x' \neq$ the unique minimal element x_0 in X (clearly, $F(x_0) = \phi \in F(G)$), $x \nleqslant x'$. Since $x = \sup\{y \in Y : y \leqslant x\}$, there is an $y \in Y$, $y \leqslant x$, such that $y \nleqslant x'$. We define an element $f \in G$ such that

$$f(t) = \begin{cases} 1, & t = y \\ 0, & t \neq y. \end{cases}$$

Then $f \in F_x$, $f \notin F_{x'}$, and it follows $F_x \nsubseteq F_{x'}$. Hence F is an injection.

Further, let $Q \in F(G)$, $Q \neq \phi$, and let $T = \{y \in Y$: there exists $g \in Q$ such that $M(g) = \{y\}\}$. It is possible to show that $(T, \leqslant)$ is a chain and according to (K1), we let $y = \sup T$ in X. Then we show that $Q = F_y$. In fact, if $f \in Q$, let f_i be components of f constructed using $M(f) = \{x_1, \ldots, x_n\}$ in the proof of 12.6.1, i.e. $f = f_1 + \cdots + f_n$. Hence, there exists i, $1 \leqslant i \leqslant n$, such that $f_i \in Q$, $M(f_i) = \{x_i\}$ and it follows $x_i \in T$, $f_i \in F_y$. Since $f \geqslant f_i$, we have $f \in F_y$ and $Q \subseteq F_y$. Conversely, let $f \in F_y$. Again, there exists i, $1 \leqslant i \leqslant n$, such that $f_i \in F_y$. If $i \neq j$, we have $0 = f_i \wedge f_j \in F_y$, a contradiction. Further, $x_i \leqslant y$ and $x_i \in Y$. Then there exists a $t << x_i$ and $t \neq y$. Thus, there exists $y' \in T$ such that $t < y' \leqslant y$. Since $y' \in T$, there exists a $g \in Q$ such that $M(g) = \{y'\}$. Now $f_i(x_i) > 0$ so there is a positive integer n such that $n(f_i(x_i)) > g(y')$ and we conclude $nf_i \geqslant g$. Hence, $nf_i \in Q$ and since Q is prime, we obtain $f_i \in Q$, $f \geqslant f_i$. Hence, $f \in Q$. Thus $F_y \subseteq Q$ and it follows $F_y = Q$. Therefore, F is a surjection and the proof is finished.

Let us observed that the theorem was generalized in Lewis, Ohm [76] in the following way. (Here a ring (with possible zero divisors) is called a *Bezout ring* if every finitely generated ideal is principal.)

12.6.2. THEOREM. *A partially ordered set* X *is a tree satisfying* (K1) *and* (K2) *(if and) only if* $X \cong \operatorname{Spec} R$ *for some Bezout ring* R.

We show their interesting proof of this theorem. At first, if a partially ordered set X is the disjoint union of partially ordered sets X_i, $i \in I$, we shall say that X is the *ordered disjoint union of the* X_i' if for all $x, y \in X$,

$x \leqslant y$ in $X \Leftrightarrow$ there exists $i \in I$ such that $x, y \in X_i$ and $x \leqslant y$ in X_i.

Now, a ring R required in 12.6.2 will be a subring of $\prod_{i \in I} R_i$ where R_i are suitable Bezout domains. Concretely, let $\{R_i\}_{i \in I}$ and $j_0 \in I$ be such that R_i is a ring and for every $i \in I' = I \setminus \{j_0\}$, R_i is an R_{j_0} - algebra via a homomorphism $\varphi_i : R_{j_0} \longrightarrow R_i$ and let

$$R = \{(r_i)_{i \in I} \in \prod_{i \in I} R_i \;:\; \varphi_i(r_{j_0}) = r_i \text{ for all but a finite number of } i \in I'\}.$$

It is possible to show without any difficulty that Spec R is order isomorphic with the order disjoint union of Spec R_i, $i \in I$.

Now, let X be a tree satisfying (K1) and (K2). Clearly, X can be written as the ordered disjoint union of trees X_i, $i \in I$, where X_i has a *unique* minimal element. Let $j_0 \in I$ and let R_{j_0} be a Bezout domain such that Spec $R_{j_0} \cong X_{j_0}$ and let K be the quotient field of R_j. Moreover, using construction from 12.6 and the proof of 8.1, we may find for every $i \in I$ a Bezout domain R_i such that Spec $R_i \cong X_i$ and $K \subseteq R_i$. Hence, R_i is an R_{j_0} - algebra via composite ring homomorphism

$$\varphi_i : R_{j_0} \hookrightarrow K \hookrightarrow R_i .$$

Therefore, the construction at the start of this proof provides a ring R such that Spec $R \cong X$.

Moreover, we show that R is a Bezout ring. Let $a^k = (a_i^k)_i \in R$, $k = 1, \ldots, n$. Then since R_{j_0} is Bezout, there is a $y_{j_0} \in R_{j_0}$ such that

$$y_{j_0} \cdot R_{j_0} = (a^1_{j_0}, \ldots, a^n_{j_0}) \cdot R_{j_0} .$$

For all but a finite number of $i \in I$ we have $\varphi_i(a^k_{j_0}) = a^k_i$; $k = 1, \ldots, n$. Clearly for such nonexceptional coordinates we have $y_i R_i = (a^1_i, \ldots, a^n_i) \cdot R_i$ for $y_i = \varphi_i(y_{j_0})$. Let $i_1, \ldots, i_m$ be the coordinates for which y_i has not be chosen. Let y_i be such that $y_{i_t} R_{i_t} = (a^1_{i_t}, \ldots, a^n_{i_t}) R_{i_t}$; $t = 1, \ldots, n$. If $y = (y_i)$; then clearly $y \in R$ and $y \cdot R = (a^1, \ldots, a^n) . R$.

The group of divisibility fields a useful tool, even in the case where we do not Theorem 8.1. The following example deals with an *essential domain*, i.e. a domain A which is an intersection of valuation domains which are quotient rings of A, and with a *v-multiplication* domain, i.e. a domain A for which the set of finitely generated v-ideals (see chapter 7) forms a group with respect to the following multiplication

$$X_v \cdot Y_v = (X. Y)_v$$

where $X \cdot Y = \{x \cdot y : x \in X, y \in Y\}$, $X, Y \subseteq G(A)$.

M. Griffin [50] has observed that every v-multiplication domain is essential and has conjectured that, in general, there exists an essential domain which is not a v-multiplication. The confirmation of this conjecture was done by Heinzer and Ohm [57] with the aid of the group of divisibility.

12.7. **EXAMPLE.** There exists an essential domain which is not a ν-multiplication.

CONSTRUCTION. To construct such a domain we introduce an auxiliary notation. If A and B are partially ordered sets such that B has infimums of any finite set of elements and if $\rho : A \longrightarrow B$ is an ordered map, we denote

$$(A, \rho, B)^{\wedge} = \{b \in B \ : \ b = \rho(a_1) \wedge \cdots \wedge \rho(a_m) \text{ for some } a_1, \ldots, a_m \in A\}.$$

If ρ is an o-epimorphism of a partially ordered sets, $(A, \rho, B)^{\wedge}$ is called an *inf-hull* of $\rho(A)$ in B, and we shall merely write $A^{\wedge}$. For a *po*-group G we denote by $S(G)$ the set of all finitely generated ν-ideals of G. Then $S(G)$ is a *po*-semigroup with

$$X_v + Y_v = (X + Y)_v,$$
$$X_v \leqslant Y_v \text{ iff } X_v \supseteq Y_v,$$

and for every $X_v, Y_v \in S(G)$ there exists $\inf(X_v, Y_v) = (X \cup Y)_v$. Then the canonical map $\rho_G(g) = (g)_v$ is an o-epimorphism, and $G^{\wedge} = S(G)$. The following proposition, the proof of which may be found in Heinzer and Ohm [57], then holds.

12.7.1. **PROPOSITION.** *Let G be a po-group, then the following conditions are equivalent.*

(1) *$S(G)$ is a group.*

(2) *There exists an l-group G' and an o-epimorphism $\rho : G \longrightarrow G'$ which is an ν-homomorphism such that $G^{\wedge}$ is a group.*

(3) *For any l-group G' and any o-epimorphism $\rho : G \longrightarrow G'$ which is an ν-homomorphism, $G^{\wedge}$ is a group.*

Assume now, that A is an essential domain, i.e. $A = \bigcap_{w \in \Omega} R_w$ where $R_w = A_{P(w)}$ for some prime ideal $P(w)$ of A. According to 2.3, $G(R_w)$ is a factor group $G(A)$ and it follows that the canonical map

$$\rho : G(A) \longrightarrow \prod_{w \in \Omega} G(R_w)$$

is an o-epimorphism and ν-homomorphism. To show that $S(G(A))$ is not a group, it suffices by 12.7.1 to prove that inf-hull $G(A)^{\wedge}$ in $\prod G(R_w)$ is not a group.

The construction sketch of such a domain is in the following.

Let k be a field and let $Y, Z, X_1, X_2, \ldots$ be indeterminates. Let

$R = k(X_1, X_2, \ldots)[Y, Z]_{(Y,Z)}$.

For each positive integer i let w_i be a valuation of a field $k(Y, Z, X_1, X_2, \ldots)$ such that

$w_i(X_i) = w_i(Y) = w_i(Z) = 1$,

$w_i(f) \geqslant 0$ for all $f \in k(X_1, X_2, \ldots)$,

$w_i(f) = \inf\{w_i(m) : m$ is a monomial occurring in $f\}$, $f \in k[Y, Z, X_1, \ldots]$.

Let $A = R \cap (\bigcap_{i \in N} R_{w_i})$. Then it is possible to show that A is an essential domain.
Let $G = G(A)$, $H = G(R)$, $Z_i = Z = G(R_{w_i})$. Then the canonical map

$\rho : G \longrightarrow H \oplus (\prod_{i \in N} Z_i)$

is an o-epimorphism and ν-homomorphism. If g is a positive element of G and $\rho(g) = (h, t_1, t_2, \ldots)$ with $h > 0$, then there exists a positive integer n such that $t_i > 0$ for all $n < i$. It follows that the infimum in $H \oplus \prod Z_i$ of finitely many positive elements of $\rho(G)$ of the form $(h, t_1, t_2, \ldots)$ with $h > 0$ also has the property that its ith coordinate is > 0 for all i is greater than some n. Let

$e = (0, 1, 1, \ldots) = \inf(\rho(w_A(Z)), \rho(w_A(Y)))$, $\rho(w_A(Y)) - e = (h, 0, 0, \ldots)$,

for some $h > 0$. Thus, the preceding consideration shows that $\rho(w_A(Y)) - e \notin G^\wedge$ even though $\rho(w_A(Y))$ and e are in $G^\wedge$. Thus, $G^\wedge$ is not a group and A is not a ν-multiplication domain.

P. M. Cohn [25] introduced the notion of a *Schreier domain*, i.e. an integrally closed integral domain A such that for each elements $a_1, a_2, c \in A$ with $c /_A a_1 \cdot a_2$ there exist $c_1, c_2 \in A$ such that $c = c_1 \cdot c_2$ and $c_1 /_A a_1$, $c_2 /_A a_2$. (Here $/_A$ is a relation of division in A.) It is easy to see that an integrally closed integral domain A is a Schreier domain if and only if $G(A)$ is a *Riesz group* (see Fuchs [38]).

For a Riesz group G, the following properties are well known :

(1) *If H is an o-ideal of G, then G/H is a Riesz group.*

(2) *The set of all o-ideals of G is a lattice with respect to the set inclusion and* $\inf\{H_1, \ldots, H_n\} = H_1 \cap \cdots \cap H_n$.

Using these facts (for example) it is possible to prove some properties of Schreier domains.

(1') *If A is a Schreier domain and S is a multiplicative system in A, then A_S is a Schreier domain.*

This follows directly from (1) and 2.3.

(2') *Let $S_1, \ldots, S_n$ be saturated multiplicative systems in a Schreier domain A. Then*

$$m_A(\overline{S_1 \cap \cdots \cap S_n}) = m_A(S_1) \cap \cdots \cap m_A(S_n)$$

where $\bar{x}$ is the saturation of X.

In fact, J. Rachunek [109] has proved that $O(G)$ is a complete lattice where for $\{H_i : i \in I\} \subseteq O(G)$, $\inf(H_i : i \in I) = [(\bigcap_{i \in I} H_i) \cap G_+]$. Then the following proposition holds.

12.8. PROPOSITION. *Let $\phi \neq \{H_i : i \in I\} \subseteq O(G(A))$, $H = \inf(H_i : i \in I)$. Then $G(A)/H \cong_0 G(A_S)$, where $S = \bigcap_{i \in I} m^{-1}(H_i)$.*

Now, to prove (2'), we obtain according to (2) and 12.8, that for $S = S_1 \cap \cdots \cap S_n$, $H_i = m_A(S_i)$, $G(A_S) \cong_0 G(A)/\inf(H_1, \ldots, H_n) = G(A)/H_1 \cap \cdots \cap H_n$ and it follows $m_A(S_1 \cap \cdots \cap S_n) = H_1 \cap \cdots \cap H_n$.

It should be observed that the opposite procedure in utilizing the group of divisibility is also possible. Namely, using a theorem in a ring theory (mostly concerning Bezout or GCD-domains) it is possible to derive some facts about *po*-groups (mostlv *l*-groups). The first step in this direction has been taken by Müller [96] in his work concerning Krull's conjecture (see 6.9). Several classical results in the ordered group theory have been derived by Mott [93] using this method.

For example, if G is an *l*-group and $\alpha \in G$, the value H_α of α is a prime *l*-ideal of G (see 1.10). This fact may be proved using Theorem 8.1.

In fact, let A be a Bezout domain such that $G(A) = G$. We may assume that $\alpha > 0$. Let $S = m_A^{-1}(H_\alpha)$, $w_A(x) = \alpha$ for some $x \in A$. Then $x \notin S$. Let P be a prime ideal of A such that $x \in P$, $P \cap S = \phi$. Then for $H = m_A(A - P)$ we have $G(A_P) \cong_0 G/H$ and since A_P is a valuation domain, H is a prime *l*-ideal of G. Moreover, $\alpha \notin H$, $H_\alpha \subseteq H$ and we have $H_\alpha = H$.

Using the approximation theorem for independent valuations we may obtain an approximation theorem for l-groups :

12.9. PROPOSITION. *Let* G *be an* l*-group,* $H_1, \ldots, H_n$ *be prime* l*-ideals of* G *such that* $H_i + H_j = G$, $i \neq j$, *and let* $\alpha_1, \ldots, \alpha_n, \beta_1, \ldots, \beta_n \in G$ *Then there exist* $\gamma, \delta_1, \ldots, \delta_n \in G$ *such that*

$$\gamma \wedge \delta_i = \gamma \wedge \beta_i = \beta_i \wedge \delta_i ,$$

$$\delta_i + H_i = \alpha_i + H_i ,$$

$$i = 1, \ldots, n .$$

P r o o f . Let A be a Bezout domain such that $G(A) = G$ and let w_i $(i = 1, \ldots, n)$ be the composition of w_A and the ith projection $G \longrightarrow G/H_i$. Then w_i is a valuation of a field K and w_i , w_j are independent for $i \neq j$. Let $b_1, \ldots, b_n \in K$ be such that $w_A(b_i) = \beta_i$. According to the approximation theorem for independent valuations there exists $a \in K$ such that

$$w_i (a - b_i) = \alpha_i + H_i , \quad i = 1, \ldots, n$$

We set $\gamma = w_A(a)$, $\delta_i = w_A(a \quad b_i)$. Since

$$(a, a - b_i) = (a - b_i , b_i) = (a, b_i) ,$$

where (a, b) is a fractional ideal of A generated by a, b we have

$$\gamma \wedge \delta_i = \delta_i \wedge \beta_i = \gamma \wedge \beta_i$$

and $\delta_i + H_i = w_i (a - b_i) = \alpha_i + H_i$.

In an l-group G a notion of a weak unit element is introduced in the following way. An element $\epsilon \in G$ is called a *weak unit element* if for any $\alpha \in G$ such that $\alpha \wedge \epsilon = 0$, $\alpha = 0$ holds. F. Šik [129] proved that for a completely regular realization $\rho : G \longrightarrow \prod_{i \in I} G_i$ of a group G an element ϵ is a weak unit element if and only if

$$Z(\epsilon) = \{ i \in I \; : \; \rho_i(\epsilon) = 0 \} = \phi$$

where ρ_i is the composition of ρ and the canonical map projection.

12.10. PROPOSITION. *Let* G *be an* *l-group and let* A *be a Bezout domain such that* $G(A) = G$. *Then in* G *there is a weak unit element if and only if the Jacobson radical* $J(A)$ *of* A *is non zero.*

P r o o f. Let $\{M_i : i \in I\}$ be the set of maximal ideals of A, $H_i = m_A(A - M_i)$. Then by 7.15, the canonical map

$$\rho : G \longrightarrow \prod_{i \in I} G/H_i$$

is a completely regular realization of G. For every $a \in A$, $a \neq 0$, we have

$$Z(w_A(a)) = \{i \in I : w_A(a) \in H_i\} = \{i \in I : a \in A - M_i\}.$$

Assume that ϵ is a weak unit element of G. Then we may supoose that $\epsilon > 0$ and $a \in A$, $w_A(a) = \epsilon$, we have $Z(\epsilon) = \phi$. Thus, $a \in J(A)$. The converse may be done analogously.

13. GENERALIZATIONS OF SEMIVALUATIONS

In an algebra and in a topology, too, the notions of valuations and their generalizations play a significant role. For example there is a lot of papers dealing with valuation theory for rings with zero divisors (Griffin [52], Fukawa [41]) with applications of valuations (and semivaluations) in topology (for the list of such papers see Wieslaw [134]) and with a theory of semivaluations for noncommutative rings (Brungs, Törner [19]. In this chapter we show the basic ideals of these methods and we will be mainly interested in generalizations of semivaluations.

We start with some generalization of a semivaluation with an l-group as a value group which was introduced by T. Nakano [102], [103] and which has a topological motivation.

It is well known (see Wieslaw [134] that almost all generalizations of valuations and semivaluations induce locally bounded topologies. Is it possible to construct some generalization of a (semi-) valuation theory such that it covers all of the locally bounded topological fields? The answer is in the affirmative and was given by Nakano.

To follow Nakano's major steps it seems to be unavoidable that we adopt a new value group. This is done in the following definition.

13.1. DEFINITION. An abelian group $(G, +)$ is called a *ν-group* if there is defined in G an associative and commutative binary operation $*$ such that

$$\alpha + (\beta * \gamma) = (\alpha + \beta) * (\alpha + \gamma)$$

for all $\alpha, \beta, \gamma \in G$.

An importance of ν-groups for our purposes follows from the fact that any ν-group is a directed *po*-group such that

$$\alpha \leqslant \beta \Leftrightarrow \alpha - \beta \in G_{-},$$

where

$G_- = \{ \alpha \in G : \text{there exists } \beta \in G \text{ such that } \alpha * \beta = 0 \}$.

In fact, according to chapter 1, we have to show that $(-G_-) + (-G_-) \subseteq -G_-$, $(-G_-) \cap (G_-) = \{0\}$ and $(-G_-) + (G_-) = G$, i.e. $-G_-$ is the set of positive elements of $(G, +, \leqslant)$, and it is routine only. For example, $\alpha * \beta \geqslant \alpha, \beta$ for any $\alpha, \beta \in G$ and it follows that G is directed.

There are several interesting examples of ν-groups.

First, let G be an l-group. Then it is easy to see that $(G, +, \vee)$ is a ν-group which coincides with the original one. Further, the set of positive real numbers R_+ is a ν-group under ordinary multiplication (= group operation) and addition (= *). It should be observed that in both cases the ν-groups are l-groups, however, not every ν-group is an l-groups. For example, let us consider the group $G = R_+ \times R_+$, where the operations are pointwise. Then G is a ν-group but G is not an l-group since (2, 1) $\wedge$ (1, 2) does not exist in G.

Moreover, a ν-group $(G, +, *)$ is an l-group such that $\alpha \vee \beta = \alpha * \beta$ for any $\alpha, \beta \in G$ if and only if $0 * 0 = 0$. (In this case we say that G is an *$l\nu$-group.*). In fact, let G be a ν-group with this property and let $\gamma \in G$ be such that $\gamma \geqslant \alpha, \beta$. Then $\alpha - \gamma, \beta - \gamma \in G_-$ and there exist $\lambda_1, \lambda_2 \in G$ such that

$$(\alpha - \gamma) * \lambda_1 = 0, \quad (\beta - \gamma) * \lambda_2 = 0.$$

Hence,

$$0 = 0 * 0 = ((\alpha - \gamma) * \lambda_1) * ((\beta - \gamma) * \lambda_2) = (\alpha * \beta - \gamma) * \lambda_1 * \lambda_2$$

and we obtain $\alpha * \beta - \gamma \in G_-$, $\gamma \geqslant \alpha * \beta$. Since $\alpha * \beta \geqslant \alpha, \beta$, we have $\alpha * \beta = \alpha \vee \beta$.

It should be observed that there is a ν-group G such that G is not a $l\nu$-group but, on the other hand, G is an l-group with respect to the ordering defined by *. In fact, let $G = R_+ \times R$ be such that

$$(\alpha, \beta) + (\alpha', \beta') = (\alpha\alpha', \beta + \beta');$$

$$(\alpha, \beta) * (\alpha', \beta') = \begin{cases} (\alpha, \beta) & \text{iff } \beta \geqslant \beta' \\ (\alpha + \alpha', \beta) & \text{iff } \beta = \beta'. \end{cases}$$

Then it is easy to see that $(G, +, *)$ is a ν-group and since $O_G = (1,0) \neq (1,0) * (1,0) =$ $= (2, 0)$, G is not a $l\nu$-group. On the other hand,

$$G_- = \{ (\alpha, \beta) \in G \ : \ (\alpha, \beta) * (\alpha', \beta') = (1, 0) \} = \{ (\alpha, \beta) \in G \ : \ \alpha \in R_+, \beta < 0 \} \cup$$
$$\cup \{ (\alpha, \beta) \in G \ : \ \beta = 0, \ 0 < \alpha < 1 \} ;$$

$$G_+ = -G_- = \{ (\alpha, \beta) \in G \ : \ \alpha \in R_+ , \beta > 0 \} \cup \{ (\alpha, 0) : \alpha > 1 \}$$

and $(G, +, \leqslant)$ is a *po*-group such that

$$(\alpha, \beta) < (\alpha', \beta') \text{ iff either } \beta < \beta' \text{ or } (\beta = \beta', \alpha < \alpha').$$

Hence, G is an *o*-group.

We are now leading to the definition of a generalized semivaluation.

13.2. **DEFINITION.** Let A be a ring. An *N-valuation* w (Nakano's valuation) of a ring A with a ν-group $(G, +, *)$ is an assignment to every element $x \in A$ a nonvoid subset $w(x)$ of G such that

(1) $w(x) = G$ if and only if $x = 0$,

(2) $w(x) * w(y) \subseteq w(x \pm y)$,

(3) $w(x) + w(y) \subseteq w(x \,.\, y)$,

for all $x, y \in A$.

It is easy to see that if w is a N-valuation with a ν-group $(G, +, *)$ then $w(x)$ is an upper class with respect to the ordering $\leqslant$ of G, i.e. $\beta \geqslant \alpha$ with $\alpha \in w(x)$ implies $\beta \in w(x)$.

The notion of a N-valuation generalizes the notion of a semivaluation with an l-group as a value group in the following sense. Let ν be a semivaluation of a field K with a value group G such that G is an l-group. As we have observed $(G, +, \vee)$ is a ν-group and if we put

$$w(x) = \nu(x) \vee G \ ; \ x \in K^*,$$
$$w(0) = G,$$

we obtain a N-valuation with the ν-group $(G, +, \vee)$. This N-valuation is called the *N-valuation associated with* ν.

Moreover, as we will see in a moment, between semivaluation ν and the N-valuation w there is a close relation. To see it we need at first to create an analogy of a semivaluation ring for N-valuation.

So let G be a ν-group and let H be a subgroup of $(G, +)$. Then H is called *N-convex* if

$$\gamma \in H \text{ and } \alpha * \beta = 0 \text{ imply } \alpha * (\beta + \gamma) \in H,$$

for every $\alpha, \beta \in G$. Since every ν-group G is a *po*-group, it is natural to consider relations between N-convex subgroups of a ν-group G and convex subgroups of a *po*-group G. Clearly any N-convex subgroup is convex. In fact, let $0 < \xi < \alpha$ for $\xi \in G$, $\alpha \in H$. Then for some $\gamma \in G$ we have $\alpha = \xi * \gamma$. Indeed, since $\alpha > \xi$. there exists γ_1 such that $(\xi - \alpha) * \gamma_1 = 0$ and it follows $\alpha = ((\xi - \alpha) * \gamma_1) + \alpha = (\xi - \alpha + \alpha) * (\gamma_1 + \alpha) = \xi * \gamma$. Analogously, there exists δ such that $\xi = 0 * \delta$. Let $\epsilon = \gamma * \delta$. Then $\alpha = 0 * \epsilon$ and let we set $\lambda = \delta - \epsilon$, $\mu = \gamma - \epsilon$. Then $\lambda * \mu = 0$ and we obtain

$$\xi = 0 * \delta = ((\gamma * \delta) - \epsilon) * \delta = (\delta - \epsilon + (0 * \epsilon)) * (\gamma - \epsilon) = (\lambda + \alpha) * \mu \in H.$$

We may show that the converse implication does not hold. Let $G = \boldsymbol{R}_+ \times \boldsymbol{R}_+$ with componentwise operations $+$ (= multiplication in $\boldsymbol{R}_+^2$) and $*$ (= addition in $\boldsymbol{R}_+^2$) and let H be a subgroup in $(G, +)$ generated by the element $(1, 2) \in G$, i.e.

$$H = \{ n \cdot (1, 2) \ : \ n \in \boldsymbol{Z} \} = \{ (1, 2^n) \ : \ n \in \boldsymbol{Z} \} .$$

Since different elements of H are incomparable with respect to the ordering of a ν-group $(G, +, *)$, H is a convex subgroup in a *po*-group $(G, +, \leqslant)$. On the other hand, we have $(2^{-1}, 2^{-1}) * (2^{-1}, 2^{-1}) = (1, 1)$ ($= 0_G$), $(1, 2) \in H$ and

$$(2^{-1}, 2^{-1}) * ((2^{-1}, 2^{-1}) + (1, 2)) = (1, 3/2) \notin H .$$

Therefore, H is not N-convex.

13.3. LEMMA. *Let G be an lv-group. Then the set of N-convex subgroups of G coincides with the set of l-ideals of the l-groups G.*

P r o o f. Let H be a N-convex subgroup of G and let $\alpha, \beta \in H$, we have

$$\alpha \vee \beta = \alpha * \beta = \alpha * (\beta + 0) \in H$$

and H is an l-ideal. Conversely, if H is an l-ideal, for $\gamma \in H$, $\alpha * \beta = 0$, we have

$$\alpha * (\beta + \gamma) = \alpha \vee (\beta + \gamma) = \alpha \vee \beta + \alpha \vee \gamma = \alpha \vee \gamma, \gamma \leqslant \alpha \vee \gamma \leqslant \gamma \vee 0,$$

and it follows $\alpha * (\beta + \gamma) \in H$. ■

13.4. LEMMA. *Let* G *be a* v*-group and let* H *be a subgroup of* $(G, +)$. *Then all the elements* γ *such that*

$$\gamma = (\lambda_1 + \gamma_1) * \ldots * (\lambda_n + \gamma_n), \quad n \in N,$$

with $\gamma_1, \ldots, \gamma_n \in H$, $\lambda_1 * \ldots * \lambda_n = 0$, *form the smallest* N*-convex subgroup* cH *of* G *containing* H.

Proof. Evidently, $H \subseteq cH$. Let S be a N-convex subgroup of G such that $H \subseteq S$ and let $\alpha, \beta \in G$. We show that

$$(\alpha + S) * (\beta + S) = (\alpha * \beta) + S.$$

In fact, let $\gamma_1, \gamma_2 \in S$. Setting $\lambda = \alpha - (\alpha * \beta)$, $\mu = \beta - (\alpha * \beta)$, we have $\lambda * \mu = 0$ and

$$(\alpha + \gamma_1) * (\beta + \gamma_2) = (\alpha * \beta)\ ((\lambda + \gamma_1) * (\mu + \gamma_2)) = (\alpha * \beta) + (\lambda * (\mu + \gamma_2 - \gamma_1)) + \gamma_1 \in$$
$$\in (\alpha * \beta) + S.$$

Moreover, let $\lambda_1, \lambda_2 \in G$ be such that $\lambda_1 * \lambda_2 = 0$ and let $\gamma_1, \gamma_2 \in S$. Then

$$(\lambda_1 + \gamma_1) * (\lambda_2 + \gamma_2) \in (\lambda_1 + S) * (\lambda_2 + S) = (\lambda_1 * \lambda_2) + S = S$$

and by induction we obtain that $cH \subseteq S$. Hence, it remains to show that cH is N-convex. Let

$$\gamma = (\lambda_1 + \gamma_1) * \ldots * (\lambda_n + \gamma_n), \quad \lambda_1 * \ldots * \lambda_n = 0,$$
$$\delta = (\mu_1 + \delta_1) * \ldots * (\mu_n + \delta_n), \quad \mu_1 * \ldots * \mu_n = 0,$$

where $\gamma_i, \delta_j \in H$. Then

$$\gamma + \delta = \mathop{*}_{\substack{1 \leq i \leq m \\ 1 \leq j \leq n}} (\lambda_i + \gamma_i + \mu_j + \delta_j) = \mathop{*}_{\substack{1 \leq i \leq m \\ 1 \leq j \leq n}} ((\lambda_i + \gamma_i + \mu_j + \gamma_i^{-1}) + \gamma_i + \delta_j)$$

where

$$\mathop{*}_{\substack{1 \leq i \leq m \\ 1 \leq j \leq n}} (\lambda_i + \gamma_i + \mu_j + \gamma_i^{-1}) = \mathop{*}_{1 \leq i \leq m} (\lambda_i + (\mathop{*}_{1 \leq j \leq n} (\gamma_i \mu_j \gamma_i^{-1}))) =$$

$$= \mathop{*}_{1 \leq i \leq m} \lambda_i = 0,$$

whence $\gamma + \delta \in cH$. Further, if we put $\lambda_i' = \lambda_i + \gamma_i - \gamma$, then $\lambda_1' * \ldots \lambda_n' = 0$ and

$0 = -\gamma \in cH$ and cH is a subgroup of G. Finally, let $\lambda * \mu = 0$. Then

$\lambda * (\mu + \gamma) = (\lambda + 0) * (\mu + \lambda_1 + \gamma_1) * \ldots * (\mu + \lambda_m + \gamma_m) \in cH$ and cH is N-convex. ∎

N-convex subgroup cH is then said to be *generated by* H. It should be observed that N-convex subgroups play the same role in the theory of ν-groups as convex subgroups in the theory of *po*-groups. In fact, we say that a map φ of a ν-group G into a ν-group H is a *homomorphism* if φ is a group homomorphism and

$$\varphi(\alpha *_G \beta) = \varphi(\alpha) *_H \varphi(\beta)$$

for all $\alpha, \beta \in G$.

The following proposition the proof of which is simple shows the principal role which play N-convex subgroups in the theory of ν-groups.

13.5. PROPOSITION. *Let* G *be a* ν*-group and let* H *be a* N*-convex subgroup of* G. *Then the factor group* G/H *is a* ν*-group with respect to the operation* $\tilde{*}$ *such that*

$$(\alpha + H) \tilde{*} (\beta + H) = (\alpha * \beta) + H$$

and the map $\alpha \longmapsto \alpha + H$ *is a homomorphism of a* ν*-group* $(G, +, *)$ *onto a* ν*-group* $(G/H, +, \tilde{*})$. *Conversely, if* φ *is a homomorphism of* G *onto a* ν*-group* H *then the* Ker φ *is a* N*-convex subgroup of* G *and* $G/\text{Ker}\ \varphi$ *is isomorphic with* H.

Let G be a ν-group and let H be a N-convex subgroup of G Then G/H is a ν-group and it follows that the multiplication $\tilde{*}$ defines an order relation $\leqslant_1$ on G/H. On the other hand, G as a ν-group is a *po*-group and on the factor group G/H we may consider the factor order relation $\leqslant_2$, i.e.

$\alpha + H \leqslant_1 \beta + H$ iff there exists $\gamma + H \in G/H$ such that

$$((-\beta + \alpha) + H) \tilde{*} (\gamma + H) = H ;$$

$\alpha + H \leqslant_2 \beta + H$ iff there exists $\gamma \in H$ such that

$$\alpha \leqslant \beta + \gamma .$$

Then $\leqslant_1 = \leqslant_2$ In fact, let $\alpha + H \leqslant_1 \beta + H$ and let $\gamma \in G$ be such that $H = ((\alpha - \beta) + H \tilde{*} (\gamma + H) = (\alpha - \beta) * \gamma + H$. Hence, $(\alpha - \beta) * \gamma = \omega \in H$ and it follows $\alpha - \beta \leqslant (\alpha - \beta) * \gamma = \omega$, $\alpha \leqslant \beta + \omega$. Thus, $\alpha + H \leqslant_2 \beta + H$. Conversely, let $\alpha + H \leqslant_2 \beta + H$, $\alpha + H \neq \beta + H$, and let $\alpha < \beta + \gamma$ for some $\gamma \in H$. Hence, $\alpha - \beta < \gamma$ and there exists $\omega \in G$ such that $\gamma = (\alpha - \beta) * \omega$ and we have $H = \gamma + H =$ $= ((\alpha - \beta) * \omega) + H = ((\alpha - \beta) + H) \tilde{*} (\omega + H)$ and $(\alpha + H) \leqslant_1 (\beta + H)$.

Now let G be a ν-group and let H_0 be the N-convex subgroup of G generated by the element $2 = 0 * 0$. Then H_0 is a *ν-subgroup* of G, i.e., if $\alpha, \beta \in H_0$ then $\alpha * \beta \in H_0$. Moreover, a N-convex subgroup H of G is a ν-subgroup if and only if $2 \in H$ and it follows that H_0 is the smallest N-convex ν-subgroup of G. If G is a $l\nu$-group then every N-convex (= convex) subgroup of G is a ν-subgroup ($2 = 0 * 0 = 0 \vee 0 = 0$). N-convex ν-subgroups of a ν-group play the same role as l-ideals of an l-group. In fact, the following simple proposition holds.

13.6. PROPOSITION. *Let* $(G, *)$ *be a ν-group and let* H *be a N-convex subgroup. Then* $(G/H, \tilde{*})$ *is $l\nu$-group if and only if* H *is a ν-subgroup of* G.

We can now define an analogy of a semivaluation ring for N-valuations. So let A be a ring and let w be a N-valuation of A with a ν-group G. Let H be a ν-subgroup of G, then

$$A_w(H) = \{x \in A \ : \ w(x) \cap H \neq \phi\}$$

is a subring of A and $A_w(H_0)$ (simply denoted by A_w) is called a *N-valuation ring of* w. (Compare with result in chapter 4.)

Let ν be a semivaluation of a field K with an l-group G as a value group and let w be the N-valuation associated with ν. Then $H_0 = \{0\}$ and

$$A_w = \{x \in K \ : \ w(x) \cap H_0 \neq \phi\} = \{x \in K \ : \ 0 \in -\nu(x) \vee G\} = A_\nu .$$

Moreover, if H is a N-convex ν-subgroup of $(G, +, \vee)$ (= l-ideal of G) then

$$A_w(H) = A_{m(H)}, \quad G(A_w(H)) \cong_0 G/H .$$

As we have mentioned above, the notion of a N-valuation is closely related to locally bounded topological fields. Concretely, for a field K and a N-valuation w with a ν-group G the sets

$$\cup_\alpha = \{x \in K \ : \ \alpha \in w(x) * G\} \ ; \ \alpha \in G ,$$

form a neighbourhood base of zero for some field topology T_w of K. Moreover, since every subset $\{\alpha\}$, $\alpha \in G$, is bounded, T_w is a locally bounded topology. On the other hand, for every locally bounded field topology T of a field K there exists a N-valuation w such that $T_w = T$ (Nakano [103], Theorem 2).

At the end of this chapter we sketch some ideas which concern groups of divisibility of non commutative rings. It should be observed that a classical concept of a group of divisibility cannot be extended directly to a not neccessarily commutative integral domain. Brungs and Törner [19] have introduced a method which enables them to associate with any ring R with a unit element and without zero-divisors a partially ordered semigroup $\widetilde{H}(R)$ which is isomorphic to the semigroup $H(A) \subseteq G(A)$ of nonzero principal ideals of A if A is a commutative domain.

In what follows let R be a ring with an unit element and without zero-divisors and let $H(R)$ be the set of principal right ideals of R. For every $x \in R$, $x \neq 0$, let $\tilde{x}$ be the mapping from $H(R)$ to $H(R)$ defined by $\tilde{x}(a \cdot R) = xaR$. If we set

$$\tilde{x} \cdot \tilde{y} = \widetilde{xy},$$

$$\tilde{x} \geqslant \tilde{y} \Leftrightarrow xaR \subseteq yaR \text{ for all } a \in R,$$

then it follows that $(\widetilde{H}/R, \leqslant, \cdot)$ is a partially ordered semigroup, where $\widetilde{H}(R) = \{\tilde{x} : x \in R, x \neq 0\}$. We set $\widetilde{R} = \{x \in R : \tilde{x} \geqslant \tilde{1}\} \cup \{0\}$. It is obvious that $\widetilde{R}$ is a subring of R. Moreover, if R is a ring such that the product of any two principal right ideals is again a principal right ideal, then $H(R)$ is a partially ordered semigroup with $aR \geqslant bR$ if and only if $aR \subseteq bR$. Then if R is *right invariant* (i.e. every right ideal is two - sided), $H(R) \cong_0 \widetilde{H}(R)$. Using $\widetilde{H}(R)$ it is possible to characterize rings which are, in some sense, valuation rings. Here we say that a ring R is a *right chain ring* if for $a, b \in R$ either $aR \subseteq bR$ or $bR \subseteq aR$ holds.

13.7. **PROPOSITION.** *The following conditions are equivalent.*

(1) $\widetilde{H}(R)$ *is totally ordered.*

(2) R *is a right chain ring such that* x *in* R, x *not in* $\widetilde{R}$ *implies* $x^{-1} \in \widetilde{R}$.

It should be observed that if R is a right chain ring it need not follow that $\tilde{H}(R)$ is totally ordered. In fact, let R_1 be a right and left chain ring, D the division ring of quotients of R_1 and let H be a totally ordered semigroup with a unit element that satisfies both cancellation laws such that the ordering of H is defined by division. Finally, let τ be a mapping from H into the semigroup $M(D)$ of monomorphisms from D to D with $\tau(h_1 h_2) = \tau(h_1) \cdot \tau(h_2)$. Then we can form the generalized power series ring

$$D\{\{H\}\} = \{\alpha = \sum_{h \in H} x_h d_h \ : \ d_h \in D \text{, supp } (\alpha) \text{ is well ordered in } H\},$$

where multiplication is defined by $x_h x_g = x_{h.g}$, $d.x_h = x_h d^{\tau(h)}$. Then the subring R of $D\{\{H\}\}$ consisting of those elements α with $\alpha_e \in R_1$ is a right chain ring where e is the unit element in H. In general it is rather complicated to investigate the order structure of $\tilde{H}(R)$. But if we set

$$R_1 = Q(x, y)_{(x)}, \quad H = (Z_+, +).$$

τ an automorphism exchanging x and y, then it is possible to show that $\tilde{H}(R)$ is not totally ordered since $\widetilde{xy^{-1}}$ and $\tilde{1}$ cannot be compared in $\tilde{H}(R)$.

Finally, we show without any proof some weak analogy of Theorem 2.3. Let R be a right invariant right chain ring with a prime ideal P and let $S = R \setminus P$, $N = \{x \in R \ : \ xa = as_a \text{ where } s_a \in S \text{ for all } a \in R\}$.

13.8. **PROPOSITION** . (1) $\tilde{H}(R_P) \cong_0 \bar{H} . \bar{S}^{-1}$, *where* $\bar{H} = \tilde{H}(R)/\tilde{N}$, $\bar{S} = \tilde{S}/\tilde{H}$.

(2) R_P *is right invariant if and only if* $N = S$.

REFERENCES

[1] Aubert, K.E.: Über Bewertungen mit halbgeordneter Wertgruppe, *Math. Ann* 127 (1954), 8 - 14.

[2] Aubert, K.E.: *Divisors of finite character*, Preprint No. 1, 1979, University of Oslo.

[3] Aubert, K.E.: Theory of *x*-ideals, *Acta Math.* 1o7 (1961), 1 - 52.

[4] Alling, N.L.: Valuation theory of meromorphic function field over open Rieman surface, *Acta Math.* 110 (1965), 79 - 95

[5] Arnold, I.: Ideale in Kommutativen Halbgruppen, *Rec. Math. Soc. Math.*, Moscow 36 (1929), 401 - 407.

[6] Banaschewski, S.D.B.: On lattice-ordered groups, *Fund. Math.* 55 (1964).

[7] Bandyopadhyaya, S.D.: Ocenki v grupach i kolcach, *Sib. Mat. Z.* 6 (1965), 1176-1180.

[8] Bandyopadhyaya, S.D.: Valuations in groups and rings, *Czech. Math. J.* 19 (94) (1969) 275 - 276.

[9] Bandyopadhyaya, S.D.: Rings with valuations, *Math. Nachr.* 44 (1970), 1 - 18.

[10] Bandyopadhyaya, S.D.: On *U*-semigroups, *Math. Nachr.* 44 (1970), 19 - 27.

[11] Birkhoff, G.: Lattice-ordered groups, *Math. Ann.* 180 (1969), 48 - 59.

[12] Borevic-Shafarevic: *Number theory*, Academic Press, New York, 1966.

[13] Bourbaki, N.: *Algèbre Commutative*, Hermann, Paris, 1965.

[14] Bourbaki, N.: *Topologie générale*, Ch. 3, Hermann, Paris, 1965.

[15] Brandal, W.: Constructing Bezout domains, *Rocky Mount. J. Math.* 6 (1976), 383 - 399.

[16] Brandal, W.: On *h*-local integral domains, *Trans. Amer. Math. Soc.* 206 (1975), 201 - 212.

[17] Brewer, J., Conrad, P., Montgomery, P.: Lattice ordered groups and conjecture for adequate domains, *Proc. A. M. S.* 43 (1974), 31 - 35.

[18] Brewer, J., Costa, D., Lady, L.: Prime ideals and localizations in commutative group rings (to appear).

[19] Brungs, H.H., Törner, R.G.: Right chain rings and generalized semigroup of divisibility, Preprint No. 425, Technische Hochschule Darmstadt 1978.

[20] Claborn, L.: Every abelain group is a class group, *Pacif. J. Math.* 18 (1966), 219 - 222.

[21] Claborn, L.: A note on class group, *Pacific J. Math.* 18 (1966), 223 - 225.

[22] Claborn, L.: A generalized approximation theorem for Dedekind domains, *Proc. A. M. S.* 18 (1967), 378 - 380.

[23] Clifford, A.H.: Partially ordered abelian groups, *Ann. Math.* 41 (1960), 465 - 473.

[24] Clifford, A.H.: Arithmetic and ideal theory of commutative semigroup, *Ann. Math.* 39 (1938), 594 - 610.

[25] Cohn, P.M.: Bezout rings and their subrings. *Proc. Cambridge Phil. Soc.* 64 (1968), 251 - 264.

[26] Cohn, P.M.: *Free rings and their relations,* Academic Press, New York, 1971.

[27] Cohn, P.M.; An invariant characterization of pseudovaluations of a field, *Proc. Cambridge Phil. Soc.* 50 (1954), 159 - 171.

[28] Cohn, P.M., Mahler, K.: On the composition of pseudo-valuations, *Nieuw. Arch. Wisk.* 3 (1953).

[29] Cohen, I.S., Kaplansky, I.: Rings with a finite number of primes, *Trans. A.M. S.* 60 (1946), 168 - 477.

[30] Connell, I.G.: On the group rings, *Can. J. Math.* 15 (1963), 650 - 685.

[31] Conrad, P.F.: *Lattice-ordered groups.* Tulune University, 1970.

[32] Conrad, P., Harvey, J., Holland, C.: The Hahn embedding theorem for Abelian lattice-ordered groups, *Trans. A. M. S.* 108 (1963), 143 - 169.

[33] Conrad, P, McAlister: The completion of lattice-ordered groups, *J. Austral. Math. Soc.* 9 (1969), 182 - 208.

[34] Dedekind, R.: Über Zerlegungen von Zahlen durch ihre grossten gemeinsame Teiler, *Math. Werke,* Chelsea 1969, 103 - 147.

[35] Dieudonne, J.: Sur la Theorie de la Divisibilité, *Bull. Soc. Math. France* 49 (1941), 1 - 12.

[36] Estes, D., Ohm, J.: Stable range in commutative rings, *J. Algebra* 7 (1967), 343 - 362.

[37] Fossum, R.: *The divisor class group of a Krull domain,* Springer-Verlag, Berlin-Heidelberg-New York 1973.

[38] Fuchs, L.: *Partially ordered algebraic systems,* Pergamon Press, New York 1963.

[39] Fuchs, L.: The generalization of the valuation theory, *Duke Math. J.* 18 (1951), 19 - 56.

[40] Fukawa, M.: An extension theorem for valuations, *J. Math. Soc. Japan* 17 (1965) 67 - 71.

[41] Fukawa, M.: On the theory of valuations, *J. Fac. Sci. Univ. Tokyo*, Ser. I, XII (1965), 57 - 79.

[42] Gilmer, R. *Multiplicative ideal theory*, Queen's papers, Lecture Notes 12, Queen's University, Kingston, 1968.

[43] Gilmer, R.: Two constructions of Prüfer domains, *J. reine angew. Math.* 239/240 (1970), 153 - 162.

[44] Gilmer, R.: An embedding theorem for HCF-rings, *Proc. Cambridge Univ.* 68 (1970), 583 - 162.

[45] Gilmer, R.: A note on semigroup rings, *Amer. Math. Montly* 75 (1969), 36 - 37.

[46] Gilmer, R., Heinzer, W.: On complete integral closure of an integral domain, *J. Austral. Math. Soc.* 6 (1966), 351 - 361.

[47] Gilmer, R., Heinzer, W.: Irreducible intersection of valuation rings, *Math. Z.* 103 (1968), 306 - 317.

[48] Gilmer, R., Heinzer, W.: Intersection of quotient rings of an integral domain, *J. Math. Kyoto Univ.* 7 (1967), 133 - 150.

[49] Gilmer, R., Parker, T.: Divisibility properties in semigroup rings, *Mich. Math.J.* 21 (1974), 65 - 86.

[50] Griffin,M.: Some results on v-multiplication rings, *Can. J. Math.* 19 (1967), 710 - 722.

[51] Griffin, M.: Rings of Krull type, *J. reine angew. Math.* 229 (1968), 1 - 27.

[52] Griffin, M.: Valuations and Prüfer rings, *Can. J. Math.* 26 (1974), 412 - 429.

[53] Hedstron, J.R.: Domains of Krull type and ideal transform, *Math. Nachr. J.* 53 (1972), 102 - 118.

[54] Helmer, O.: Divisibility properties of integral functions, *Duke Math. J.* 6 (1940), 345 - 356.

[55] Heinzer, W.: J-Noetherian integral domains with 1 in the stable range, *Proc. A. M. S.* 19 (1968), 1369 - 1372.

[56] Heinzer, W.: Some remarks on complete integral closure, *J. Austral. Math. Soc.* 9 (1969), 310 - 314.

[57] Heinzer, W., Ohm, J.: Locally noetherian commutative rings, *Trans. A. M. S.* 158 (1971), 273 - 284.

[58] Heinzer, W., Ohm, J.: Defining families for integral domains of real finite character, *Can J. Math.* 24 (1972), 1170 - 1177.

[59] Heinzer, W., Ohm, J.: An essential ring which is not a ν-multiplication ring, *Can. J. Math.* 25 (1973), 856 - 861.

[60] Heinzer, W., Ohm, J.: Noetherian intersection of integral domains, *Trans. A. M. S.* 167 (1972), 291 - 308.

[61] Hill, P.: On the complete integral closure of a domain, *Proc. A. M. S.* 36 (1972), 26 - 30.

[62] Henriksen, M.: Some remarks on elementary divisor rings II, *Mich. Math. J.* (1955/56), 156 - 163.

[63] Huckaba, J.: Extension of pseudo-valuations, *Pac. J. Math.* 29 (1969), 295 - 302.

[64] Huckaba, J.: Some results on pseudo-valuations, *Duke Math. J.* 37 (1970), 1 - 9.

[65] Jaffard, P.: *Les systémes d' Ideaux,* Dunod, Paris, 1960.

[66] Jaffard, P.: Contribution à la thèorie des groupes ordonnés, *J. Math. Pures Appl.* 32 (1953), 203 - 280.

[67] Jaffard, P.: Extension des groupes réticulés et applications, *Publ. Sci. Unive. Ager.* Ser. A-1 (1954), 197 - 222.

[68] Jaffard, P.: Un contre-example concernant les groupes de divisibilité, *C. R. Acad. Sci. Paris* 243 (1956), 1264 - 1268.

[69] Jaffard, P.: Corps demie-value, *C. R. Acad. Sci. Paris* 231 (1950), Ser. A-B, 1401 - 1403.

[70] Jaffard, P.: La notion de valuation, *Enseignement Math.* 40 (1955), 5 - 26.

[71] Jaffard,P.: Solution d'un probleme de Krull, *Bull. Soc. Math. France* 85 (1961), 127 - 135.

[72] Kowalsky, H.J.: Beiträge zur topologischen Algebra, *Math. Nachr.* 9 (1953), 261 - 268.

[73] Levi, J.: Contribution to the theory of ordered groups, *Proc. Indiana Acad. Sci.* 17 (1943), 199 - 201.

[74] Lorenzen, P.: Abstrakte Begründung der multiplicative Idealtheorie, *Math. Z.* 45 (1939), 533 - 553.

[75] Lewis, W.J.: The spectrum of a ring as a partially ordered set, *J. Algebra* 25 (1973), 419 - 434.

[76] Lewis, W.J., Ohm, J.: The ordering of Spec R, *Can. J. Math.* 28 (2976), 820 - 835.

[77] Madell, R.L.: Embedding of topological lattice-ordered groups, *Trans. A. M. S.* 146 (1969), 447 - 455.

[78] Krull, W.: Algemeine Bewertrungtheorie, *J. reine angew. Math.* 117 (1931), 160 - 196.

[79] Krull, W.: Beiträge zur arithmetik kommutativen Integritätsbereich I, II, *Math. Z.* 41 (1936), 544 - 577, 665 - 679.

[80] Krull, W.:Zur Theorie der Bewertungen mit nichtarchimedisch geordneter Wertgruppe und der nichtarchimedisch geordneten Körper, *Colloq. d' Algebre Superieure,* Bruxelles, 1956, 45 - 77.

[81] Močkoř, J.: Semi-valuations and *d*-groups, to appear, in *Czech. Math. J.*

[82] Močkoř, J.: On *o*-ideals of group of divisibility, *Czech. Math. J.* 31 (1981), 390 - 403.

[83] Močkoř, J.: A realization of *d*-groups, *Czech. Math. J.* 27 (1977), 296 - 312.

[84] Močkoř, J.: Topological groups of divisibility, *Colloq. Math.* XXXIX (1978), 301 - 311.

[85] Močkoř, J.: Prüfer *d*-groups, *Czech. Math. J.* 28 (1978), 127 - 139.

[86] Močkoř, J.: A realization of groups of divisibility, *Com. Math. Univ. St. Pauli,* Tokyo, 26 (1977), 61 - 75.

[87] Močkoř, J.: A note on approximation theorem, *Arch. Math.* 2 (1979), 107 - 118.

[88] Močkoř, J.: The completion of Prüfer domains, *Proc. A. M. S.* 67 (1977), 1 - 10.

[89] Močkoř, J.: The completion of valued fields and nonstandard models, *Com. Math. Univ. St. Pauli,* Tokyo, 30 (1981), 1 - 16.

[90] Mockoř, J.: The group of divisibility of $\hat{Z}$, to appear in *Arch. Math.*

[91] Mott, J.L.: Non spliting sequence of value groups, *Proc. A. M. S.* 44 (1979), 39 - 42.

[92] Mott, J.L.: Convex directed subgroups of a group of divisibility, *Can. J. Math.* 26 (1974), 532 - 542.

[93] Mott, J.L.: The group of divisibility and its applications, 1972, Conference on Com. Algebra, Kansas, Springer-Verlag, 1973.

[94] Mott, J.L.: The group of divisibility of Rees rings, *Math. Japonical* 20 (1975), 85 - 87.

[95] Mott, J.L., Schexnayder, M.: Exact sequences of semi-value groups, *J. reine angew. Math.* 283/284 (1976), 388 - 401.

[96] Müller, D.: Verbandsgruppen und Durchschnitte endlich vieler Bewertungsringe, *Math. Z.* 77 (1961), 45 - 62.

[97] Nagata, M.: *Local rings,* Interscience, New York, 1963.

[98] Nagata, M.: A remark on the unique factorization theorem, *J. Math. Soc. Japan* 9 (1957), 143 - 145.

[99] Nakano, T.: A theorem on lattice ordered groups and its applications to the valuation theory, *Math. Z.* 83 (1964), 140 - 146.

[100] Nakano, T.: Rings and partly ordered systems, *Math. Z.* 99 (1967), 355 - 376.

[101] Nakano, T.: Integrally closed integral domains, *Com. Math. Univ. St. Pauli, Tokyo* 18 (1970), 53 - 59.

[102] Nakano, T.: A generalized valuations and its value group, *Com. Math. Univ, St. Pauli Tokyo* 12 (1964), 1 - 22.

[103] Nakano, T.: On the locally bounded fields, *Com. Math. Univ. St. Pauli Tokyo* 9 (1961).

[104] Nakayama, T.: On Krull conjecture concerning completely integrally closed integrity domains I, II, III, *Proc. Imp. Acad. Tokyo* 18 (1942), 185 - 187, 233 - 236, *Proc. Japan Acad.* 22 (1946), 249 - 250.

[105] Ohm, J.: Semi-valuations and group of divisibility, *Can. J. Math.* 21 (1969) 576 - 591.

[106] Ohm, J.: Some counterexamples related to integral closure in $D[[X]]$, *Trans. A. M. S.* 122 (1966), 322 - 333.

[107] Parker, T.: The semigroup ring, Dissertation, Florida State Univ., 1973.

[108] Prüfer, H.: Untersuchunge über die Teilbarkeitsegenschaften in Körpern, *J. reine angew. Math.* 168 (1932), 1 - 36.

[109] Rachunek, J.: Directed convex subgroups of ordered groups, *Acta Univ. Palac. Olom. Fac. R. Nat.* 41 (1973), 39 - 46.

[110] Rachunek, J.: Prime subgroups of ordered groups, *Czech. Math. J.* 24 (1974), 541 - 551.

[111] Ribenboim, P.: Sur les groupes totalement ordonnés et l'arithmétique des anneaux du valuation, *Summa Brasil. Math.* 4 (1958).

[112] Ribenboim, P.: Modules dur un anneaux de Dedekind, *Summa Brasil. Math.* 3 (1952).

[113] Ribenboim, P.: Sur quelques constructions de groupes réticulés et l'equivalence logique entre l'affinement de filtres et d'ordres, *Summa Brasil. Math.* 4 (1958).

[114] Ribenboim, P.: Anneaux normaux rèels à caractére fini, *Summa Brasil. Math.* 3 (1956).

[115] Ribenboim, P.: Le thèoréme d'approximation pour les valuations de Krull, *Math. Z.* 68 (1957), 1 - 18.

[116] Ribenboim, P.: Un thèoréme de réalisation de groupes réticulés, *Pac. J. Math.* 10 (1960), 305 - 308.

[117] Sheldon, P.: Two counterexamples involving complete integral closure in finite dimensional Prüfer domains, *J. Algebra* 32 (1974), 152 - 172.

[118] Sheldon, P.: Prime ideals in GCD-domains, *Can. J. Math.* (to appear).

[119] Sheldon, P.: A counter example to a conjecture of Heinzer (to appear).

[120] Sheldon, P.: How changing $D[[X]]$ changes its quotient field, *Trans. A. M. S.* 159 (1971), 223 - 244.

[121] Skula, L.: Divisorentheorie einer Halbgruppe, *Math. Z.* 114 (1970), 113 - 120.

[122] Skula, L.: Forsetzung stetiger Homomorphismem von δ-Halbgruppen, *J. reine angew. Math.* 261 (1973), 71 - 87.

[123] Skula, L.: Maximale δ- und δ_1- Kategorien, *J. reine angew. Math.* 274/275 (1975), 287 - 298.

[124] Skula, L.: On c-semigroups, *Acta arithmetica* XXXI (1976), 247 - 257.

[125] Skula, L.: On δ_n-semigroups, *Arch. Math.* (Brno) XVII (1981), 43 - 52.

[126] Sik, F.: Closed and open sets in topologies introduced by lattice ordered vector groups, *Czech. J. Math.* 23 (1973), 139 - 150.

[127] Sik, F.: Estructure y realizaciones de grupos reticulados, *Memom. Fac.Cie. Univ. Habana* vol. 1, Ser. Mat. I, II, No. 3 (1964), 1 - 11, 11 - 29, III No. 4 (1966), 1 - 20, IV No. 7 (1968), 19 - 44.

[128] Sik, F.: Verbandsgruppen deren Komponentenverbands Kompakt erzeugt ist, *Scripta Fac. Sci. Nat. Univ. Brun. Arch. Math.* 3 (1971), 101 - 121.

[129] Sik, F.: Struktur und Realisierungen von Verbandsgruppen *v*, *Math. Nachr.* 33 (1967), 221 - 229.

[130] Sik, F.: Archimedische kompakt erzeugte Verbandsgruppen, *Math. Nach.* 38 (1968), 323 - 340.

[131] Smarda, B.: Topologies in *l*-groups, *Arch. Math.* 3 (1967), 69 - 81.

[132] Smarda, B.: Some types of topological *l*-groups, *Spisy prirodoved. fak. Univ. J.E. Purkyne,* Brno, 507 (1969), 341 - 352,

[133] Wiegand, S.: Locally-maximal Bezout domains, *Proc. A. M. S.* 47 (1975), 10 - 14.

[134] Wieslaw, W.: On topological fields, *Colloq. Math.* 29 (1974), 119 - 146.

[135] Yakabe, I.: On semi-valuations, *Mem. Fac. Sci. Kyushu Univ.* Ser. A 17 (1963), 10 - 28.

[136] Zariski, O., Samuel, P.: Commutative algebra II, Princeton 1960.

[137] Zakon, E.: A new variant of non-standard analysis, Victoria Symposium on Nonstandard Analysis, *Lecture Notes in Mathematics*, Vol. 369, springer Verlag, Berlin, 1974.

SUBJECT INDEX

A

additive semivaluation 14
approximation theorem 56
archimedean *d*-group 34

B

basic element 10
basis 10
Bezout domain 14
bounded element 151

C

canonical *d*-valuation of a field 29
c-characteristic 128
class group 123
compatible elements 49
compatible system from a localization 60
completely regular *l*-realization 93
completely integrally closed *m*-ring 34
composite of rings 112
Conrad's (*F*) - conditions 101
convex subgroup 5

D

d-convex subgroup 28
dense subgroup 116
d-epimorphism 26
d-group 22
d-group associated with a domain 22
d-group related to an *m*-ring 26
δ-group 122
δ_1-group 122
d-homomorphism 26
directed group 5
disjoint elements 6
d-isomorphism 26

divisor class group 123
d-realization 43
d-valuation of a field 29
d-valuation of a *d*-group 36

E

element integral over 36
essential domain 157
exact addition 25

F

family compatible with ρ 89
finitely dense subgroup 126
fractional *m*-ideal 43

G

group algebra 96
group of divisibility 13
GCD-domain 14

I

inf-hull 158
integrally closed *m*-ring 36
invertible *m*-ideal 43
irreducible *d*-realizator 92
irreducible realization 10
irredundant representation 92
I-realization 93

K

Krull domain 120

L

lex - exact sequence 11
lex - extension 11
lexicographic sum 11
lexicographically spliting 11
l-group 5
l-homomorphism 6
l-ideal 6

local *d*-group 25
locally-bounded representation 135
locally-compact representation 136
locally-Noetherian domain 153
Lorenzen r-group 109

M

m-ideal 26
m-ring 26

N

normal group 120

O

o-epimorphism 6
o-exact sequence 11
o-extension 11
o-group 5
o-homomorphism 6
o-ideal 6
o-isomorphism 6
o-product 8
o-sum 8

P

po-group 1 5
polar 8
positive cone 5
presheaf 60
prime *d*-convex subgroup 28
prime filter 7
prime *m*-ideal 26
prime *o*-ideal 31
Prüfer *d*-group 43
pseudo Bezout domain 14
Π' - realization 93

Q

quasidiscrete category 56
quasisheaf 61

R

r-closed *po*-group 108
realization 8
realizator 11
reduced *l*-realization 88
r-ideal 88
ring of Krull type 101
r-system 88

S

Schreier domain 159
semiclosed group 13
semivaluation 8
semivalue group 14
semivaluering 13
sheaf 60
site 60
solution of compatible system 60
spliting sequence 11
strong system of generators 123

T

theory of divisors of a domain 121
theory of divisors of a *po*-group 122
theory of quasidivisors 116
tl-isomorphism 130
t-prime ideal 7
t-realizator 130
tree 154
t-system 107

U

UF - group 122
UF - property 118
ultraproduct 21

V

valuation *m*-ring 36
value group 13
value of element 10

ν-homomorphism 88
ν-multiplication domain 157
ν-system 88

W

weak unit element 161
well centred overring 30
w-system 90

9 789027 715395